Hands-On Geospatial Analysis with R and QGIS

A beginner's guide to manipulating, managing, and analyzing spatial data using R and QGIS 3.2.2

Shammunul Islam

BIRMINGHAM - MUMBAI

Hands-On Geospatial Analysis with R and QGIS

Commissioning Editor: Richa Tripathi
Acquisition Editor: Divya Poojari
Content Development Editor: Ishita Vora
Technical Editor: Snehal Dalmet
Copy Editor: Safis Editing
Project Coordinator: Namrata Swetta
Proofreader: Safis Editing
Indexer: Pratik Shirodkar
Graphics: Jisha Chirayil
Production Coordinator: Arvindkumar Gupta

First published: November 2018

Production reference: 1301118

Published by Packt Publishing Ltd.
Livery Place
35 Livery Street
Birmingham
B3 2PB, UK.

ISBN 978-1-78899-167-4

www.packtpub.com

mapt.io

Mapt is an online digital library that gives you full access to over 5,000 books and videos, as well as industry leading tools to help you plan your personal development and advance your career. For more information, please visit our website.

Why subscribe?

- Spend less time learning and more time coding with practical eBooks and Videos from over 4,000 industry professionals

- Improve your learning with Skill Plans built especially for you

- Get a free eBook or video every month

- Mapt is fully searchable

- Copy and paste, print, and bookmark content

Packt.com

Did you know that Packt offers eBook versions of every book published, with PDF and ePub files available? You can upgrade to the eBook version at www.packt.com and as a print book customer, you are entitled to a discount on the eBook copy. Get in touch with us at customercare@packtpub.com for more details.

At www.packt.com, you can also read a collection of free technical articles, sign up for a range of free newsletters, and receive exclusive discounts and offers on Packt books and eBooks.

Contributors

About the author

Shammunul Islam is a consulting spatial data scientist at the Institute of Remote Sensing, Jahangirnagar University. His guidance is being applied toward the development of an adaptation tracking mechanism for a UNDP project in Bangladesh. He has provided data science training to the executives of Shwapno, the largest retail brand in Bangladesh. Mr. Islam has developed applications for automating statistical and econometric analysis for a variety of data sources, ranging from weather stations to socio-economic surveys. He has also consulted as a statistician for a number of surveys. He completed his MA in Climate and Society from Columbia University, New York, in 2014 on a full scholarship, before which he completed an honors degree in statistics and a master's degree in development studies.

About the reviewers

Ana-Cornelia BADEA is a professor at the Faculty of Geodesy, Technical University of Civil Engineering, Bucharest. She defended habilitation in the field of geodetic engineering, and is a director of the Engineering Geodetic Measurements and Spatial Data Infrastructures Research Center. She is a UTCB representative at FIG. Her research interests focus primarily on modern geospatial data acquisition technologies, 3D modeling, GIS analysis, GIS-BIM integration, project management, concepts of urban cadastre, cadastral GIS applications, mobile mapping, and web GIS applications. She is the author and co-author of over 90 scientific papers at national and international conferences, as well as 10 books. She is president of the editorial board of the Journal of Geodesy, Cartography, and Cadastre, and the Union of Romanian surveyors, and is involved in numerous international editorial committees. She is a member of the ASRO ISO TC 359 committee on geospatial data standardization and is involved in project evaluation for national and international calls.

Brad Hamson is a spatial analyst and developer in the Seattle area whose professional interests include spatial data engineering, systems engineering, remote sensing, data science, and designing geospatial software applications. He is currently a graduate student pursuing a Master of Science degree in engineering management from the School of Engineering and Applied Science at George Washington University. He holds a Bachelor of Science degree in geography and environmental planning from Towson University, with a focus on geographic information sciences. Brad has extensive experience designing, implementing, and operating enterprise geographic information systems and solutions using proprietary and open source technologies. His specialist areas include spatial analysis using Python, system architecture and design, software development, spatial database design, data visualization, cartography, and graphic design.

Chima Obi is the lead geospatial analyst at AGERPoint Inc. His areas of expertise include processing lidar data, feature extraction from raster files, data visualization, big data analytics, and Python and R programming, as well as exploring other open source geospatial tools. He attained his bachelor's degree in soil science from the Federal University of Technology Owerri, Nigeria, in 2010. He then moved to the United States, where he obtained his master's degree in environmental science and obtained a certificate in geospatial information systems in 2016.

Prior to working at AGERPoint, he worked as a geospatial analyst at the West Virginia District of Highways throughout 2015 and 2016. He has extensive experience in the analysis of geospatial data.

I would like to express my gratitude to my friends and family, most importantly to my wife, for their wonderful encouragement and support. Also, my biggest thanks go to Packt Publishing for choosing me to be part of this awesome book review.

Packt is searching for authors like you

If you're interested in becoming an author for Packt, please visit `authors.packtpub.com` and apply today. We have worked with thousands of developers and tech professionals, just like you, to help them share their insight with the global tech community. You can make a general application, apply for a specific hot topic that we are recruiting an author for, or submit your own idea.

Table of Contents

Preface

Where is something happening? Are there similarities between different areas with respect to an attribute of interest? Which area is most susceptible to a particular hazard? These and many other questions can be answered if you take location into account in your analysis. Location plays an important role and has critical implications for many policy decisions regarding environment, biodiversity, and socio-economy. This area is increasingly being studied by researchers and practitioners from many disciplines. In particular, in the realization of **Sustainable Development Goals (SDGs)**, **Geographic Information Systems (GIS)**, and **remote sensing (RS)**, data can play a pivotal role.

R and QGIS are two examples of open source software that can be used free of charge for working with spatial data. By using them, we can answer many of our questions regarding location. For the last couple of years, R, a language originally intended for statisticians, has also been used as GIS software. We can readily call any spatial package in R and apply it to our data. QGIS is very powerful GIS software that enables users to perform many complex spatial tasks. QGIS and R both have a very strong user community and, by combining these two according to their relative advantages, you can perform very sophisticated and complex spatial analysis tasks.

This book covers both R and QGIS, unlike the other books on the market. Assuming you have zero, or rudimentary, knowledge of GIS and RS, this book will have transformed you from a beginner to an intermediate user of GIS and RS by the time you finish it. This book guides you from the initial step of setting up the software, to spatial analysis, geostatistics, and applying different models for landslide susceptibility mapping by providing hands-on examples, code, and screenshots. After reading this book, you should be able to generalize the examples to your spatial problems and create susceptibility maps using machine learning algorithms.

Who this book is for

This book is great for geographers, environmental scientists, statisticians, and professionals who deal with spatial data. If you want to learn how to handle GIS and RS data, then this book is for you. Basic knowledge of R and QGIS would be helpful, but is not necessary.

What this book covers

Chapter 1, *Setting Up R and QGIS Environments for Geospatial Tasks*, shows how to set up the R and QGIS environments necessary for this book. The basics of R programming are covered, and you are introduced to the interface of QGIS.

Chapter 2, *Fundamentals of GIS Using R and QGIS*, details the different ways that spatial data is handled by R and QGIS. You are introduced to the steps that need to be followed to set up different projection systems and re-project data in this software. Packages such as sp, maptools, rgeos, sf, ggplot2, ggmap, and tmap in R are covered, showing how spatial data can be imported, exported, and visualized with the R engine. This chapter also shows how to do the same tasks with QGIS, with the help of detailed descriptions and screenshots. You will learn how to visualize quantitative and qualitative data in both R and QGIS.

Chapter 3, *Creating Geospatial Data*, provides a detailed overview of how to create geospatial data. This chapter will shed light on how vector and raster data is stored and how you can create point data, line data, and polygon data. Using QGIS, you will also be introduced to the digitization of maps.

Chapter 4, *Working with Geospatial Data*, explains how to query data for information extraction, how to use different joins, how to dissolve polygons, how to use buffering, and more. R and QGIS are both used to accomplish these tasks.

Chapter 5, *Remote Sensing Using R and QGIS*, begins with the basics of RS. The steps required to load and visualize remote sensing in R and QGIS are followed by band arithmetic, stacking and unstacking raster images, and other basic operations with RS data.

Chapter 6, *Point Pattern Analysis*, starts with the basic terminology of **point pattern process (PPP)** such as points, events, marks, windows, the spatial point pattern, and the spatial point process. It then explains how to use R to create R objects. You are then introduced to the PPP analysis for spatial randomness checking using quadrat testing, G-function, K-function and L-function, and others.

Chapter 7, *Spatial Analysis*, introduces readers to testing and modeling autocorrelation, fitting generalized linear models, and geostatistics. Checking the spatial autocorrelation of data using Moran's I is covered here, followed by spatial regression and a generalized linear model. Spatial interpolation and the basics of geostatistics are also discussed here.

Chapter 8, *GRASS, Graphical Modelers, and Web Mapping,* focuses on some more open source software, GRASS GIS, which can be used with QGIS. The chapter explains how to set up GRASS GIS and perform GRASS operations. Automating tasks using the graphical modeler is also covered. You will also learn how to make web maps inside QGIS.

Chapter 9, *Classification of Remote Sensing Images,* covers the basics of remote sensing image classification using QGIS 3.2.2. Supervised classification using the SCP plugin of QGIS is used to show how you can classify landsat images.

Chapter 10, *Landslide Susceptibility Mapping,* is a case study-based chapter where you are introduced to the different steps needed to make landslide susceptibility maps. Using the historical data of landslide events in Bangladesh, this chapter provides a step-by-step guide to the process of creating a landslide susceptibility map. In doing so, R and QGIS are used together. Logistic regression and decision-tree-based algorithms are used to fit models, and the accuracy of those models are then computed.

To get the most out of this book

Basic knowledge of mathematics, such as addition and subtraction, is sufficient to follow this book.

If you are working with spatial data or plan to work with spatial data, you will benefit most by generalizing the examples in this book to your research question.

Download the example code files

You can download the example code files for this book from your account at www.packt.com. If you purchased this book elsewhere, you can visit www.packt.com/support and register to have the files emailed directly to you.

You can download the code files by following these steps:

1. Log in or register at www.packt.com.
2. Select the **SUPPORT** tab.
3. Click on **Code Downloads & Errata**.
4. Enter the name of the book in the **Search** box and follow the onscreen instructions.

Once the file is downloaded, please make sure that you unzip or extract the folder using the latest version of:

- WinRAR/7-Zip for Windows
- Zipeg/iZip/UnRarX for Mac
- 7-Zip/PeaZip for Linux

The code bundle for the book is also hosted on GitHub at `https://github.com/PacktPublishing/Hands-On-Geospatial-Analysis-with-R-and-QGIS`. In case there's an update to the code, it will be updated on the existing GitHub repository.

We also have other code bundles from our rich catalog of books and videos available at `https://github.com/PacktPublishing/`. Check them out!

Download the color images

We also provide a PDF file that has color images of the screenshots/diagrams used in this book. You can download it here: `http://www.packtpub.com/sites/default/files/downloads/9781788991674_ColorImages.pdf`.

Conventions used

There are a number of text conventions used throughout this book.

CodeInText: Indicates code words in text, database table names, folder names, filenames, file extensions, pathnames, dummy URLs, user input, and Twitter handles. Here is an example: "Mount the downloaded `WebStorm-10*.dmg` disk image file as another disk in your system."

A block of code is set as follows:

```
jan_price = c(10, 20, 30)
increase = c(1, 2, 3)
mar_price = jan_price + increase
```

When we wish to draw your attention to a particular part of a code block, the relevant lines or items are set in bold:

```
# 6th element in the Polygons slot of map_bd
sixth_element = map_bd@polygons[[6]]
# make it succinct with max.level = 2 in str() for the 6th element of the
bd@Polygons
str(sixth_element, max.level = 2)
```

Bold: Indicates a new term, an important word, or words that you see onscreen. For example, words in menus or dialog boxes appear in the text like this. Here is an example: "Select **System info** from the **Administration** panel."

Warnings or important notes appear like this.

Tips and tricks appear like this.

Get in touch

Feedback from our readers is always welcome.

General feedback: If you have questions about any aspect of this book, mention the book title in the subject of your message and email us at customercare@packtpub.com.

Errata: Although we have taken every care to ensure the accuracy of our content, mistakes do happen. If you have found a mistake in this book, we would be grateful if you would report this to us. Please visit www.packt.com/submit-errata, selecting your book, clicking on the Errata Submission Form link, and entering the details.

Piracy: If you come across any illegal copies of our works in any form on the Internet, we would be grateful if you would provide us with the location address or website name. Please contact us at copyright@packt.com with a link to the material.

If you are interested in becoming an author: If there is a topic that you have expertise in and you are interested in either writing or contributing to a book, please visit authors.packtpub.com.

Reviews

Please leave a review. Once you have read and used this book, why not leave a review on the site that you purchased it from? Potential readers can then see and use your unbiased opinion to make purchase decisions, we at Packt can understand what you think about our products, and our authors can see your feedback on their book. Thank you!

For more information about Packt, please visit `packt.com`.

1
Setting Up R and QGIS Environments for Geospatial Tasks

This chapter will walk its readers through the different stages of setting up the R and QGIS environments. R and QGIS are both free and open source software that can be used for various geospatial tasks. R benefits from more than 10,000 packages developed by its community, and QGIS also benefits from a number of plugins that are available to QGIS users. QGIS can complement R, and vice versa, for the conduct of many sophisticated geospatial tasks, and many statistical and machine learning algorithms can be very easily applied using R with the help of QGIS.

The first segment of the book starts by discussing how to install R and getting to know its environment. That is followed by data types in R, and different operations in R, and then getting acquainted with writing functions and plotting. The second segment consists of installing QGIS, learning the QGIS environment, and getting help in QGIS.

The following topics are to be covered in this chapter:

- Installing R
- Basic data types and the data structure in R
- Looping, functions, and apply family in R
- Plotting in R
- Installing QGIS
- Getting to know the QGIS environment.

Installing R

R is an open source programming language and software used for statistical computing and graphics, which has benefited greatly from the continuous contributions of its user community. Graphics in R are of very high quality, and, although it was not primarily developed for GIS purposes, with the development of packages such as **ggmap**, **tmap**, **sf**, **raster**, **sp**, and so on, R can work as a GIS environment itself. Furthermore, R codes can be written inside QGIS and we can also work on QGIS inside R using the **RQGIS** package.

We will now install R with the help of snapshots of each of the step-by-step instructions provided. The following steps have been implemented in Microsoft Windows and should also be applicable to Mac or other platforms with a little tweaking. There are no specific requirements for computer configuration, but any modern desktop or laptop will be sufficient to run the examples provided in this book.

Download R software from the following site and click on **download R**: https://www.r-project.org/.

Now we need to select a CRAN mirror; we will use the first link to download R.

Now we will need to click on **Download R for Windows**:

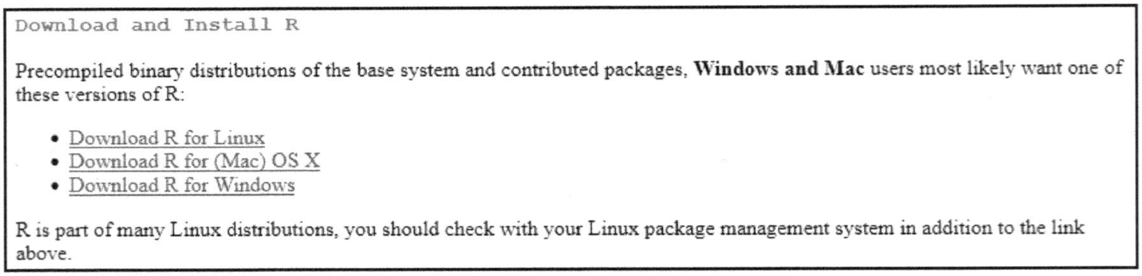

Download and Install R

Precompiled binary distributions of the base system and contributed packages, **Windows and Mac** users most likely want one of these versions of R:

- Download R for Linux
- Download R for (Mac) OS X
- Download R for Windows

R is part of many Linux distributions, you should check with your Linux package management system in addition to the link above.

Click **install R for the first time**, as we can see from the following screenshot:

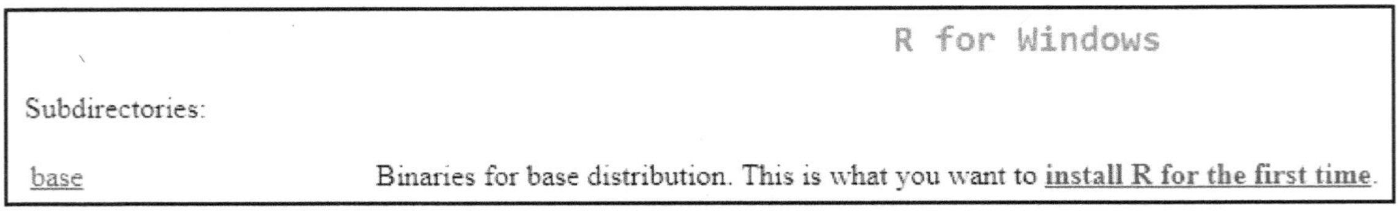

R for Windows

Subdirectories:

base   Binaries for base distribution. This is what you want to <u>install R for the first time</u>.

Now we just need to double-click the `.exe` file that we have downloaded and continue to click to accept all the defaults to complete the download of R. After we have installed R, we need to open it, and it will look similar to the following screenshot. For this installation process, a 64-bit computer is being used, but if you are using a 32-bit computer, your R windows will reflect that:

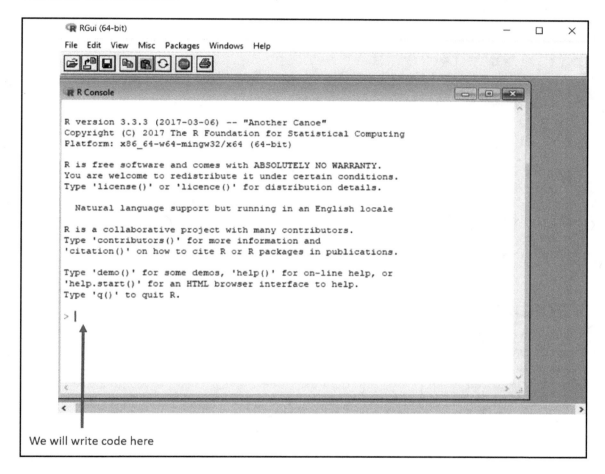

We are finally ready to rock and roll using R. But before that, we need a little bit more familiarity with R, or perhaps we need a refresher.

Basic data types and data structures in R

Before we start delving deep into R for geospatial analysis, we need to have a good understanding of how R handles and stores different types of data. We also need to know how to undertake different operations on that data.

Basic data types in R

There are three main data types in R, and they are as follows:

- Numerics
- Logical or Boolean
- Character

Numerics are any numbers with decimal values; thus, 21.0 and 21.1 are both numerics. We can use addition, subtraction, multiplication, division, and so on, with these numerics. Interestingly, R also considers integer numbers to be numerics. **Logical** or **Boolean** data consists of TRUE and FALSE; they are mainly used for different comparisons. The **character** variable consists of text, such as the name of something. We write character values in R by putting our character values inside " ", or double quotes.

Variable

Just before digging any deeper, we need to know how to assign values to any variable. So, what is a variable? It's like a container, which holds different value(s) of different types (or the same type). When assigning multiple values to any variable, we write the variable name to the left, followed by an <- or = and then the value. So, if we want to assign 2 to a variable x, we can write either of the two:

```
x <- 2
```

or

```
x = 2
```

I find the latter convenient, although the R community prefers to use the former – my suggestion is to use one which you find more convenient.

Data structures in R

The data structures in R are as follows:

- Vectors
- Matrices
- Arrays
- Data frames
- Lists
- Factors

Vectors

Vectors are used to store single or multiple values of similar data types in a variable and are considered to be one-dimensional arrays. That means that the x variable we just defined is a vector. If we want to create a vector with multiple numeric values, we assign as before with one additional rule: we put all the values inside c() and separate all the values with , except the last value. Let's look at an example:

```
val = c(1, 2, 3, 4, 5, 6)
```

What happens if we mix different data types such as both numerics and characters? It works! (A variable's name is arbitrarily named as val, but you can name your variable anything that you feel appropriate, anything!) Except in some cases, such as variable names, shouldn't start with any special character:

```
x = c(1, 2.0, 3.0, 4, 5, "Hello", "OK")
```

What we have just learned about storing data of the same types doesn't seem to be true then, right? Well, not exactly. What R does behind the scenes is that it tries to convert all the values mentioned for the x variable to the same type. As it can't convert Hello and OK to numeric types, for conformity it converts all the numeric values 1, 2.0, 3.0, 4, and 5 to character values: that is, "1", "2.0", "3.0", "4", and "5", and adds two more values, "Hello" and "OK", and assigns all these character values to x. We can check the class (data type) of a variable in R with class(variable_name), and let's confirm that x is indeed a character variable:

```
class(x)
```

We will see that the R window will show the following output:

```
[1] "character"
```

We can also label vectors or give names to different values according to our need. Suppose we want to assign temperature values recorded at different times to a variable with a recorded time as a label. We can do so using this code:

```
temperature = c(morning = 20, before_noon = 23, after_noon = 25, evening =
22, night =   18)
```

Basic operations with vector

Suppose the prices of three commodities, namely potatoes, rice, and oil were $10, $20, and $30 respectively in January 2018, denoted by the vector `jan_price`, and the prices of all these three elements increased by $1, $2, and $3 respectively in March 2018, denoted by the vector `increase`. Then, we can add two vectors `mar_price` and `increase` to get the new price as follows:

```
jan_price = c(10, 20, 30)
increase = c(1, 2, 3)
mar_price = jan_price + increase
```

To see the contents of `mar_price`, we just need to write it and then press *Enter*:

```
mar_price
```

We now see that `mar_price` is updated as expected:

```
[1] 11 22 33
```

Similarly, we can subtract and multiply. Remember that R uses element-wise computation, meaning that if we multiply two vectors which are of the same size, the first element of the first vector will be multiplied by the first element of the second vector, and the second element of the second vector will be multiplied by the second element of the second vector, and as such:

```
x = c(10, 20, 30)
y = c(1, 2, 3)
x * y
```

The result of this multiplication is this:

```
[1] 10 40 90
```

If we multiply a vector with multiple values by a single value, that latter value multiplies every single element of the vector separately. This is demonstrated in the following example:

```
x * 2
```

We can see the output of the preceding command as follows:

```
[1] 20 40 60
```

As a vector does element-wise computation, if we check for any condition, the condition will be checked for each element. Thus, if we want to know which values in x are greater than 15:

```
x > 15
```

As the second and third elements satisfy this condition of being greater than 15, we see TRUE for these positions and FALSE for the first position as follows:

```
[1] FALSE TRUE TRUE
```

Indexing in R or the first element of any data type starts with 1; thus, the third or fourth element in R can be accessed with index 3 or 4. We need to access any particular index of a variable with a variable name followed by the index inside []. Thus, the third element of x can be accessed as follows:

```
x[3]
```

By pressing *Enter* after x[3], we see that the third element of x is this:

```
30
```

If we want to select all items but the third one, we need to use – in the following way:

```
x[-3]
```

We now see that x has all of the elements except the third one:

```
[1] 10 20
```

Matrix

Suppose, we also have the prices of these three items for the month of June as follows:

```
june_price = c(20, 25, 33)
```

Now if we want to stack all these three months in a single variable, we can't use vectors anymore; we need a new data structure. One of the data structures to rescue in this case is the matrix. A matrix is basically a two-dimensional array of data elements with a number of rows and columns fixed. Like a vector, a matrix can also contain just one type of element; a mix of two types is not allowed. To combine these three vectors with every row corresponding to a particular month's prices of different items and every column corresponding to prices of different items in a particular month, what we can do is first combine these three vectors inside a `matrix()` command, followed by a comma and `nrow` = 3, indicating the fact that there are three different items (for example, items are arranged row-wise and months are arranged column-wise).

```
all_prices = matrix(c(jan_price, mar_price, june_price), nrow= 3)
all_prices
```

The `all_prices` data frame will look like the following:

```
     [,1] [,2] [,3]
[1,]  10   11   20
[2,]  20   22   25
[3,]  30   33   33
```

Now suppose we change our mind and want to arrange this with the items displayed column-wise and the prices displayed row-wise; that is, the first row corresponds to the prices of different items in a particular month and the first column corresponds to the first month's (January's) prices of different items, with that arrangement continuing for every other row and column. We can do so very easily by mentioning `byrow` = TRUE inside the matrix. `byrow` = TRUE arranges the values of vectors row-wise. It arranges the matrix by aligning all the elements row-wise allowing for its dimensions:

```
all_prices2 = matrix(c(jan_price, mar_price, june_price), nrow= 3, byrow =
TRUE)
all_prices2
```

The output will look like the following:

```
     [,1] [,2] [,3]
[1,]  10   20   30
[2,]  11   22   33
[3,]  20   25   33
```

We can see that here `jan_price` is considered as the first row, `mar_price` as the second row, and `june_price` as the third row in `all_prices2`.

Array

Arrays are also like matrices, but they allow us to have more than two dimensions. The `all_prices2` row has prices of different items for January, March, and June 2018. Now, suppose we also want to record prices for 2017. We can do so by using `array()` and in this case we want to add two 3x3 matrices where the first one corresponds to 2018 and the latter matrix corresponds to 2017. In `array(m, n, p)`, m and n stand for the dimensions of the matrix and p stands for how many matrices we want to store.

In the following example, we define six vectors for three different months for two different years. Now we create an array by combining six different vectors using `c()` and by using them inside `array()` as inputs as follows:

```
# Create six vectors
jan_2018 = c(10, 11, 20)
mar_2018 = c(20, 22, 25)
june_2018 = c(30, 33, 33)
jan_2017 = c(10, 10, 17)
mar_2017 = c(18, 23, 21)
june_2017 = c(25, 31, 35)

# Now combine these vectors into array
combined = array(c(jan_2018, mar_2018, june_2018, jan_2017, mar_2017,
june_2017),dim = c(3,3,2))
combined
```

We can now see that we have two matrices of 3 x 3 dimensions, as in the output as follows:

```
, , 1

     [,1] [,2] [,3]
[1,]   10   20   30
[2,]   11   22   33
[3,]   20   25   33

, , 2

     [,1] [,2] [,3]
[1,]   10   18   25
[2,]   10   23   31
[3,]   17   21   35
```

Data frames

Data frames are like matrices, except for the one additional advantage that we can now have a mix of different element types in a data frame. For example, we can now store both numeric and character elements in this data structure. Now, we can also put the names of different food items along with their prices in different months to be stored in a data frame. First, define a variable with the names of different food items:

```
items = c("potato", "rice", "oil")
```

We can define a data frame using data.frame as follows:

```
all_prices3 = data.frame(items, jan_price, mar_price, june_price)
all_prices3
```

The data frame all_prices3 looks like the following:

```
  items jan_price mar_price june_price
1 potato       10        11         20
2   rice       20        22         25
3    oil       30        33         33
```

Accessing elements in a data frame can be done by using either [[]] or $. To select all the values of mar_price or the second column, we can do either of the two methods provided as follows:

```
all_prices3$mar_price
```

This gives the values of the mar_price column of the all_prices3 data frame:

```
[1] 11 22 33
```

Similarly, there is the following:

```
all_prices3[["mar_price"]]
```

We now find the same output as we found by using the $ sign:

```
[1] 11 22 33
```

We can also use [] to access a data frame. In this case, we can utilize both the row and column dimensions to access an element (or elements) using the row index indicated by the number before, and the column index indicated by the number after. For example, if we wanted to access the second row and third column of all_prices3, we would write this:

```
all_prices3[2, 3]
```

This gives the following output:

```
[1] 22
```

Here, for simplicity, we will drop items column from `all_prices3` using – and rename the new variable as `all_prices4` and we can define this value in a new vector `pen` as follows:

```
all_prices4 = all_prices3[-1]

all_prices4
```

We can now see that the `items` column is dropped from the `all_prices4` data frame:

```
  jan_price mar_price june_price
1        10        11         20
2        20        22         25
3        30        33         33
```

We can add a row using `rbind()`. Now we define a new numerical vector that contains the price of the `pen` vector for January, March, and June, and we can add this row using `rbind()`:

```
pen = c(3, 4, 3.5)
all_prices4 = rbind(all_prices4, pen)
all_prices4
```

Now we see from the following output that a new observation is added as a new row:

```
  jan_price mar_price june_price
1        10        11       20.0
2        20        22       25.0
3        30        33       33.0
4         3         4        3.5
```

We can add a column using `cbind()`. Now, suppose we also have information on the prices of `potato`, `rice`, `oil`, and `pen` for August as given in the vector `aug_price`:

```
aug_price = c(22, 24, 31, 5)
```

We can now use `cbind()` to add `aug_price` as a new column to `all_prices4`:

```
all_prices4 = cbind(all_prices4, aug_price)
all_prices4
```

Now `all_prices4` has a new column `aug_price` added to it:

```
  jan_price mar_price june_price aug_price
1        10        11       20.0        22
2        20        22       25.0        24
3        30        33       33.0        31
4         3         4        3.5         5
```

Lists

Now, items `jan_price` and `mar_price` have four elements, whereas `june_price` has three elements. So, we can't use a data frame in this case to store all of these values in a single variable. Instead, we can use **lists**. Using lists, we can get almost all the advantages of a data frame in addition to its capacity for storing different sets of elements (columns in the case of data frames) with different lengths:

```
all_prices_list2 = list(items, jan_price, mar_price, june_price)
all_prices_list2
```

We can now see that `all_prices_list2` has a different structure than that of a data frame:

```
[[1]]
[1] "potato" "rice"   "oil"

[[2]]
[1] 10 20 30

[[3]]
[1] 11 22 33

[[4]]
[1] 20 25 33
```

Accessing list elements can be done by either using `[]` or `[[]]` where the former gives back a list and the latter gives back element(s) in its original data type. We can get the values of `jan_price` in the following way:

```
all_prices_list2[2]
```

Using `[]`, we are returned with the second element of `all_prices_list2` as a list again:

```
[[1]]
[1] 10 20 30 15
```

Note that, by using `[]`, what we get back is another list and we can't use different mathematical operations on it directly.

```
class(all_prices_list2[2])
```

We can see, as follows, that the class of `all_prices_list2` is a list:

```
[1] "list"
```

We can get this data in original data types (that is, a numeric vector) by using `[[]]` instead of `[]`:

```
all_prices_list2[[2]]
```

Now, we get the second element of the list as a vector:

```
[1] 10 20 30 15
```

We can see that it is numeric and we can check further to confirm that it is numeric:

```
class(all_prices_list2[[2]])
```

The following result confirms that it is indeed a numeric vector:

```
[1] "numeric"
```

We can also create categorical variables with `factor()`.

Suppose we have a numeric vector x and we want to convert it to a factor, we can do so by following the code as shown as follows:

```
x = c(1, 2, 3)
x = factor(x)
class(x)
```

Factor

We now see that the class is a `factor`, as we can see in the following output:

```
[1] "factor"
```

Now, we can also look at the internal structure of this vector x, using `str()` as follows:

```
str(x)
```

We now see that it converts 1, 2, and 3 to factors:

```
[1]  Factor w/ 3 levels "1", "2", "3": 1 2 3
```

Looping, functions, and apply family in R

Looping allows us to do repetitive task in a couple of lines of code, saving us much effort and time. Functions allow us to write a block of instructions that could be modified to work according to the way they are being called. Combining the power of looping, functions, and apply family in R allows us to loop through the elements of a data type, or similar, and apply a function or use a block of instructions on each of these.

Looping in R

Suppose we want to loop through all the values of the `aug_price` column inside `all_prices4` and square them and return them. We can do so in the following way:

```
jan = all_prices4$jan_price
for(price in jan){
    print(price^2)
}
```

This prints a square of all the prices in January as follows:

```
[1]  100
[1]  400
[1]  900
[1]  9
```

Functions in R

We can also achieve the previous result by using a function. Let's name this function `square`:

```
square = function(data){
    for(price in data){
        print(price^2)
    }
}
```

Now call the function as follows:

```
square(all_prices4$jan_price)
```

The following output also shows the squared price of `jan_price`:

```
[1] 100
[1] 400
[1] 900
[1] 9
```

Now suppose we want to have the ability to take elements to any power, not just `square`. We can attain it by making a little tweak to the function:

```
power_function = function(data, power){
  for(price in data){
    print(price^power)
  }
}
```

Now suppose we want to take the power of 4 for the price in June, we can do the following:

```
power_function(all_prices4$june_price, 4)
```

We can see that the `june_price` column is taken to the fourth power as follows:

```
[1] 160000
[1] 390625
[1] 1185921
[1] 150.0625
```

Apply family – lapply, sapply, apply, tapply

We discuss apply family here, which allows us not to have to write loops and reduces our workload. We will discuss four functions under this family: apply, lapply, sapply, and tapply.

apply

apply works on arrays or matrices and gives us an easier way to compute something row-wise or column-wise. For the apply() function, this row- or column-wise consideration is denoted by a margin. The apply() function takes the following form: apply(data, margin, function). This data has to be an array or a matrix, and the margin can be either 1 or 2, where 1 stands for a row-wise operation and 2 stands for a column-wise operation. We will work with the matrix all_prices, which has the following structure:

```
     [,1] [,2] [,3]
[1,]   10   11   20
[2,]   20   22   25
[3,]   30   33   33
```

Here, we have a record of prices of three different items in three different months (January, March, and June), where a row represents the prices of an item in three different months and a column represents the prices of three different items in any single month. Now, if we want to know which item's price fluctuated most over these three months, we would have to compute a standard deviation row-wise for each row. We can do this very easily using margin = 1 in apply().

```
apply(all_prices, 1, sd)
```

We can see the standard deviation for these three items as follows:

```
[1] 5.507571 2.516611 1.732051
```

Now suppose we want to know the month-wise total cost of all three items. As every column corresponds to different months, we can apply apply() with margin = 2 and a function mean to achieve this:

```
apply(all_prices, 2, sum)
```

This gives the sum for all three months in a vector:

```
[1] 60 66 78
```

We see that the total prices were the highest in June (the third column), totaling 78.

> Note that the function that we use inside apply() has to be without (). We just need to write its name without parentheses.

lapply

In the previously mentioned power_function() function, we had to use a for loop to loop through all the values of the june_price column of the all_prices4 data frame. lapply allows us to define a function (or use an already existing function) over all the elements of a list or vector and it returns a list. Let's redefine power_function() to allow for the computation of different powers on elements and then use lapply to loop through each element of a list or vector and take the power of each of these elements on every iteration of the loop. lapply() has the following format:

```
lapply(data, function, arguments_of_the_function)

power_function2 = function(data, power){
    data^power
}
lapply(all_prices4$june_price, power_function2, 4)
```

As we saw in the last output, all the prices of june_price are taken to the fourth power and are returned as a list:

```
[[1]]
[1] 160000

[[2]]
[1] 390625

[[3]]
[1] 1185921

[[4]]
[1] 150.0625
```

> What we get in return is a list. We can use unlist() to get a simple vector for our convenience.

```
unlist(lapply(all_prices4$june_price, power_function2, 4))
```

Now we are returned the fourth power of the `june_price` column as a vector.

```
[1]  160000.0000  390625.0000 1185921.0000      150.0625
```

Now we will again work with a **combined** array, which has the prices of different items in three different months each for 2017 and 2018. Do you remember the structure of it? It looked like this:

```
, , 1

     [,1] [,2] [,3]
[1,]  10   20   30
[2,]  11   22   33
[3,]  20   25   33

, , 2

     [,1] [,2] [,3]
[1,]  10   18   25
[2,]  10   23   31
[3,]  17   21   35
```

Here, the first matrix corresponds to prices for 2017 and the second matrix corresponds to 2018. We will now recreate this array to become a list of matrices in the following way:

```
combined2 = list(matrix(c(jan_2018, mar_2018, june_2018), nrow = 3),
  matrix(c(jan_2017, mar_2017, june_2017), nrow = 3))
combined2
```

This returns us the following list of matrices:

```
[[1]]
     [,1] [,2] [,3]
[1,]  10   20   30
[2,]  11   22   33
[3,]  20   25   33

[[2]]
     [,1] [,2] [,3]
[1,]  10   18   25
[2,]  10   23   31
[3,]  17   21   35
```

Now, if we want the prices for March for both 2017 and 2018, we can use `lapply()` in the following way:

```
lapply(combined2, "[", 2,)
```

```
[[1]]
[1] 11 22 33

[[2]]
[1] 10 23 31
```

So, what this has done is selected the second row from each list:

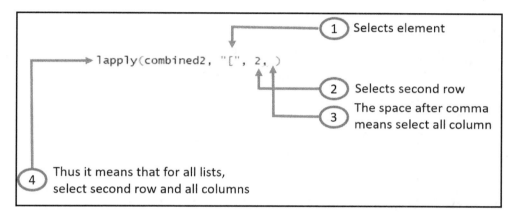

Now we can modify it further to select a column, row, or any element according to our needs.

    `lapply()` can be used with data frames, lists, and vectors.

sapply

What we have got by using `unlist(lapply(data, function, arguments_of_the_function))` can be obtained simply by using `sapply(data, function, arguments_of_the_function)`.

```
sapply(all_prices4$june_price, power_function2, 4)
```

We are returned with a vector again as follows:

```
[1]   160000.0000   390625.0000 1185921.0000      150.0625
```

Now let's go back to the example of the `all_prices3` data frame. We can see this from the screenshot that follows:

```
  items jan_price mar_price june_price
1 potato        10        11         20
2   rice        20        22         25
3    oil        30        33         33
```

tapply

Now, suppose instead of prices for 2018 only, we have prices for these items for 2017, 2016, and 2015 as well. This new data frame is defined as follows:

```
all_prices = data.frame(items = rep(c("potato", "rice", "oil"), 4),
          jan_price = c(10, 20, 30, 10, 18, 25, 9, 17, 24, 9, 19,27),
          mar_price = c(11, 22, 33, 13, 25, 32, 12, 21, 33, 15, 27,39),
          june_price = c(20, 25, 33, 21, 24, 40, 17, 22, 27, 13, 18,23)
                                  )
all_prices
```

The output for the preceding lines of code can be seen as follows:

```
   items jan_price mar_price june_price
1  potato        10        11         20
2    rice        20        22         25
3     oil        30        33         33
4  potato        10        13         21
5    rice        18        25         24
6     oil        25        32         40
7  potato         9        12         17
8    rice        17        21         22
9     oil        24        33         27
10 potato         9        15         13
11   rice        19        27         18
12    oil        27        39         23
```

Now suppose we want to take the mean price of different items for very March in all years. We can do this by using `tapply(numerical_variable, categorical_variable, function)`. So, we will need to convert the items column of the `all_prices` data frame to a categorical variable to take the mean price.

```
tapply(all_prices$mar_price, factor(all_prices$items), mean)
```

This gives us a mean March price for `oil`, `potato`, and `rice` in all years, as follows:

```
  oil potato   rice
34.25  12.75  23.75
```

Note the use of `factor()` to convert the items column to a factor variable.

There are other `apply` functions, but that's it for now, folks. We will introduce new functions as and when it will be necessary as we proceed to new chapters for geospatial analysis.

To install a new package, we need to write `install.packages("package_name")`, and to use any package, we need to write `load.packages("package_name")`.

Plotting in R

We can make a simple plot using the `plot()` function of R. Now we will simulate 50 values from a normal distribution using `rnorm()` and assign these to x and similarly generate and assign 50 normally distributed values to y. We can plot these values in the following way:

```
x = rnorm(50)
y = rnorm(50)
# pch = 19 stands for filled dot
plot(x, y, pch = 19, col = 'blue')
```

This gives us the following scatterplot with blue-colored filled dots as symbols for each data point:

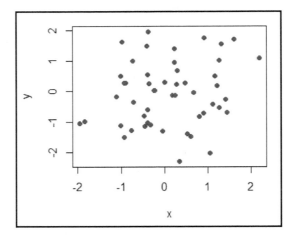

We can also generate a line plot type of graph by using `type = "l"` inside `plot()`.

Now we will briefly look at a very strong graphical library called `ggplot2` developed by Hadley Wickham. Remember, the `all_prices` data frame? If you don't, let's have another look at that:

```
str(all_prices)
```

We see that it has 12 rows and four columns, it has three numeric variables and one factor variable:

```
data.frame':    12 obs. of  4 variables:
 $ items     : Factor w/ 3 levels "oil","potato",..: 2 3 1 2 3 1 2 3 1 2 ...
 $ jan_price : num  10 20 30 10 18 25 9 17 24 9 ...
 $ mar_price : num  11 22 33 13 25 32 12 21 33 15 ...
 $ june_price: num  20 25 33 21 24 40 17 22 27 13 ...
```

We first need to install and then load the `ggplot2` package:

```
install.packages("ggplot2")
library(ggplot2)
```

TIP

In any R session, if we want to use an R package, we need to load it using `library()`. But once loaded, we don't need to load it any further to use any of the functions inside the package.

Now we need to define the data frame we want to use inside the `ggplot()` command, and inside this command, after the data frame name, we need to write `aes()`, which stands for **aesthetics**. Inside this `aes()`, we define the *x* axis variable and the *y* axis variable. So, if we want to plot the prices of different items in January against these items, we can do the following:

```
ggplot(all_prices, aes(x = items, y = jan_price)) +
geom_point()
```

Now we see the plot as follows:

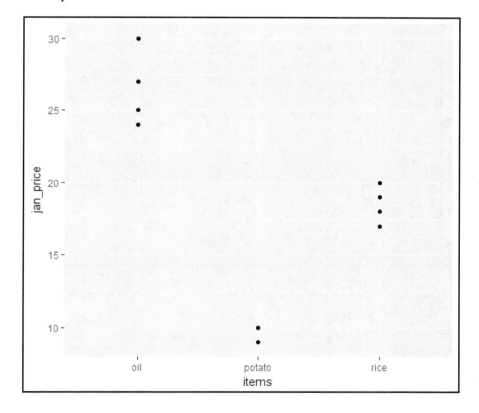

We can also compute and mark the mean price in January of these different items over all the years under consideration using `stat = "summary"` and `fun.y = "mean"`. We will just need to add another layer, `geom_point()`, and mention these arguments inside this:

```
ggplot(all_prices, aes(x = items, y = jan_price)) +
  geom_point() +
  geom_point(stat = "summary", fun.y = "mean", colour = "red", size = 3)
```

The following screenshot shows that along with data points, the mean values for each item are marked as red:

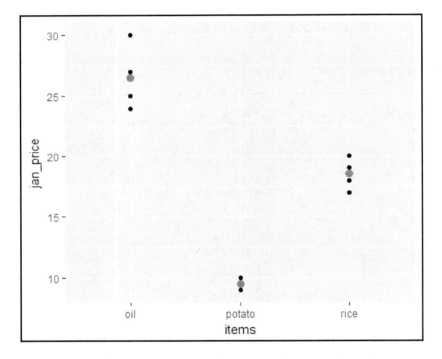

We can also plot the price of January against the price of June and make separate plots for each of the items using `facet_grid(. ~ items)`:

```
ggplot(all_prices, aes(x = jan_price, y = june_price)) +
  geom_point() +
  facet_grid(. ~ items)
```

As a result, we see a scatterplot for three different items as follows:

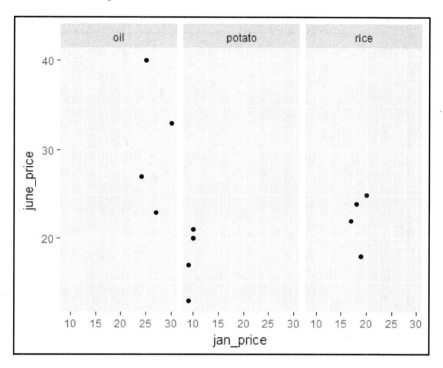

We can also add a linear model fit using a `stat_smooth()` layer:

```
ggplot(all_prices, aes(x = jan_price, y = june_price)) +
geom_point() +
facet_grid(. ~ items) +
# se = TRUE inside stat_smooth() shows confidence interval
stat_smooth(method = "lm", se = TRUE, col = "red")
```

The preceding code gives a linear model fit and a 95% confidence interval along with the scatterplot:

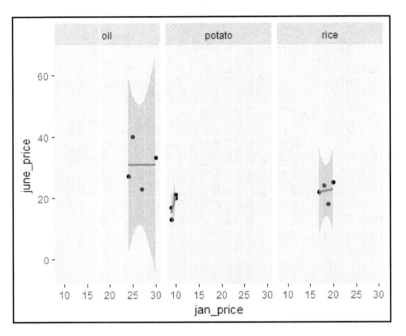

We get this weird-looking confidence interval for the `oil` price and the `rice` price, as there are very few points available.

We can do so many more things, and we have so many other things to cover in this book that we will not be covering any more plotting functionalities here. But we will explain many other aspects of plotting as and when appropriate when dealing with spatial data in upcoming chapters. I have also listed books to refer to for a deeper understanding of R in the *Further reading* section.

Installing QGIS

QGIS is a free and open source **geographic information system (GIS)** that we can use for various spatial data management and analysis tasks for different fields, such as geography, environmental science, disaster management, urban planning, climate science, and many other fields that use spatial data. The strength of QGIS lies in the fact that it is an open source platform coupled with different plugins available for computing different tasks.

QGIS can be installed in different operating systems such as Windows, Mac, Linux, Android, and so on. QGIS can be installed from the following site:

`http://download.osgeo.org/qgis/win64/`

After going to the previously mentioned website, we will scroll down and click on QGIS-OSGeo4W-3.2.2-1-Setup-x86_64.exe to download QGIS 3.2.2-1 (or click on the installer relevant to the operating system you are using):

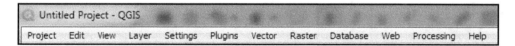

QGIS-OSGeo4W-3.2.2-1-Setup-x86.exe	18-Aug-2018 12:27	413M
QGIS-OSGeo4W-3.2.2-1-Setup-x86.exe.md5sum	18-Aug-2018 11:41	69
QGIS-OSGeo4W-3.2.2-1-Setup-x86_64.exe	18-Aug-2018 12:47	474M
QGIS-OSGeo4W-3.2.2-1-Setup-x86_64.exe.md5sum	18-Aug-2018 12:10	72

Now, if you are a Mac user, you need to install the **Geospatial Data Abstraction Library** (**GDAL**) framework and the matplotlib module of Python before installing QGIS. You can do so from this address: `http://www.kyngchaos.com/software/qgis`

Getting to know the QGIS environment

The QGIS desktop is used to display, analyze, and to do different design formatting with data. The QGIS desktop has four main components: a **Menu bar**, **Tool Bars**, **Panels**, and **Map Display**. The **Menu bar** is the top section and appears as follows:

Just under this, we have **Tool Bars**, which look like this:

We then have the **Panels** section on the left side, which is composed of these parts: **Browser** and **Layers**. The **Browser** gives us different options for data connection and working with layers. **Layers** shows all the vector and raster files that we can load to **QGIS**:

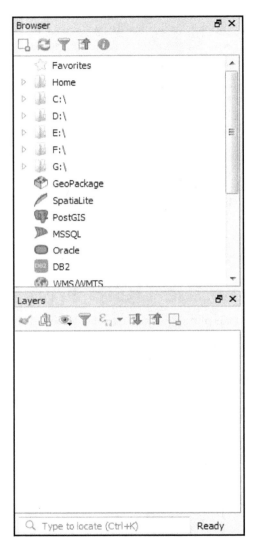

We also have **Map Display**, which shows us the map outputs:

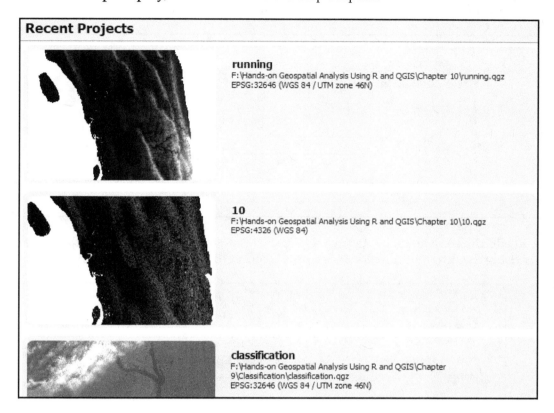

In the **Map Display** section, as shown in the preceding screenshot, we see some of the projects the author has been working on; in your case, if you are starting afresh, this section will be blank at first.

Using QGIS, we can complete many geospatial data management and spatial data analysis tasks. The following is a screenshot of some of the useful sections of QGIS:

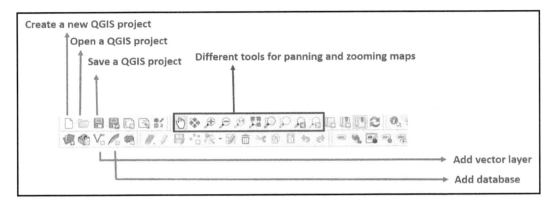

We can add different spatial data such as vector layers, raster layers, and also database layers using the different functionalities provided in QGIS: **Layer** | **Add Layer** | ...:

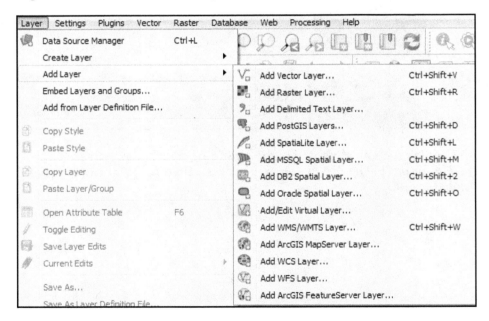

We will now add a vector file (shapefile) in QGIS to illustrate how it is being done in this GIS software. Suppose we want to add the file BGD_adm3.shp to our QGIS environment; we can do this by following these steps:

1. Click on **Add Vector Layer...** under **Add Layer**, which is under **Layer**:

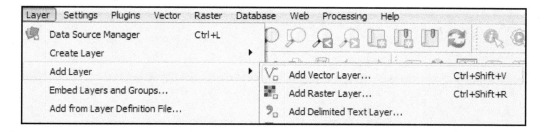

2. We click on the indicated rectangular shape to browse to the folder where the shapefile we want to add is located (in the **Data** folder under **Chapter 1**):

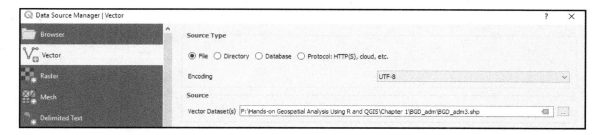

3. Now we browse to **BGD_adm3.shp** under the **Data** folder of **Chapter 1**, select it, and then press Close:

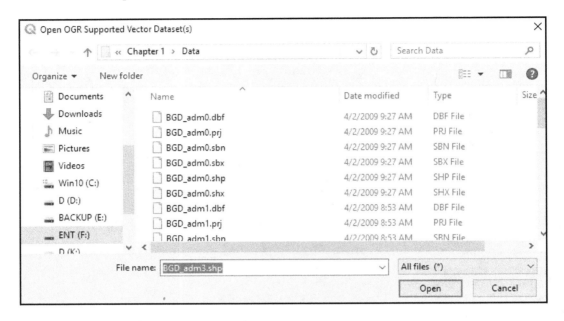

4. Now click **Add** and then click **Close**. Now, we will see a map of the districts of Bangladesh in the **Map Display** as follows:

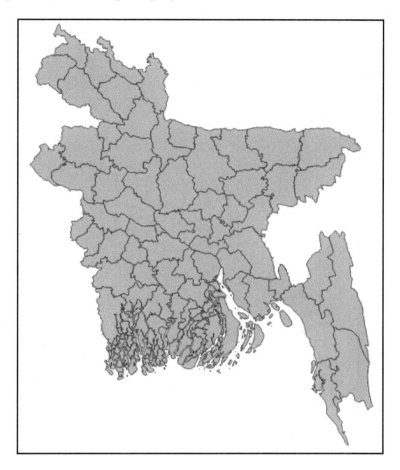

Now we will look at one more important aspect of vector data: its attribute table. An attribute table contains information about the shape of points, lines, and polygon features, or mainly the geometry of features, in addition to any other information associated with those features. This information is recorded in a tabular form, where each row represents a record and each column corresponds to field or a feature. We can access this table by right-clicking on the **BGD_adm3** layer in the **Layers** panel of QGIS and then by left-clicking on **Open Attribute Table**:

Now we will see the attribute table associated with this shapefile:

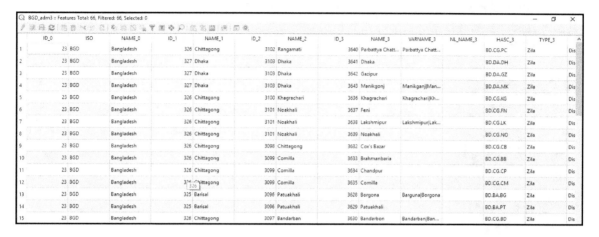

Similar to adding a vector layer, we can add a raster layer (or layers) and other database layer(s) to the QGIS environment, which we will look at more in-depth as we proceed further in this book.

QGIS has a number of plugins that are add-ons that increase the functionality of QGIS. We can click on **Plugin** in the **Menu Bar** and then click on **Manage and Install Plugins** to install new plugins, as shown in this screenshot:

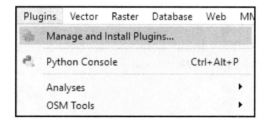

Now we will see a list of available plugins. Select the plugin you want to install, and then click on **Install plugin**. When the installation is finished, click **Close**:

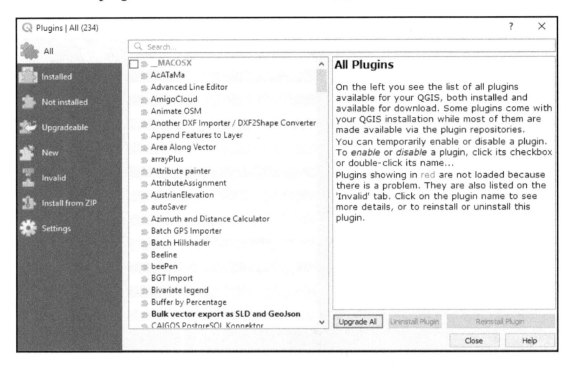

In the next chapter, we will look at the basics of GIS and **remote sensing** (**RS**) and we will explore further how R and QGIS handle them and how we can use these two software for basic geospatial data loading and visualization.

Summary

In this chapter, we have learned how to download and install R and QGIS. We started with the installation of R, following which we also saw the various data types in R and how to work with these in R. Later in this chapter, we studied the programming aspects of R and also learned to use and apply loops and functions. Additionally, we saw how to visualize data in R using the `ggplot2` package. Finally, we also learned about installing QGIS and also plugins, and we briefly studied the QGIS desktop.

We have only just scratched the surface of the many functionalities of R and QGIS. We have yet to touch upon working with spatial data, creating a spatial database, conducting spatial data analysis, and so on, which we will be introducing in Chapter 2, *Fundamentals of GIS Using R and QGIS*. Working with spatial data in R and QGIS requires us to know about the basics of GIS and how spatial data is being handled by R and QGIS, which we will be discussing in detail in Chapter 2. So, let's jump in!

Questions

If we have followed this chapter closely, by now, we should be able to answer the following questions:

- How do users install R?
- What are the basic data types in R?
- How can users work with these different types of data in R?
- How would users loop through a number of values in R?
- How would users write a function in R?
- What are lapply, sapply, and tapply and how are they used in R?
- How is data plotted in R?
- How do users install QGIS?
- How can users explore an attribute table in QGIS?
- How can users add vector (and raster) data in QGIS?
- How can users install plugins in QGIS?

Further reading

To get a good idea about the various aspects of data management and writing functions in R, please refer to *R Cookbook* by Paul Teetor and *Advanced R* by Hadley Wickham. If you are looking for a thorough introduction to QGIS, please refer to the books *QGIS2 Cookbook* by A Mandel et al and *Mastering QGIS* by K Menke et al.

Fundamentals of GIS Using R and QGIS

In this chapter, we will learn the basics of how **geographic information system (GIS)** data is handled in R and QGIS and we will also learn about the different facilities provided by these two software for spatial data visualization. The focus will be on understanding the basics of vector data in the light of R and QGIS. Starting with the basics of GIS, projection systems, and the loading and visualization of these data in R and QGIS, we will be covering all of these with hands-on examples. We will start with the basics of spatial (here, vector) data, which will be followed by basic vector data loading and visualization in R, and lastly cover basic vector data loading and visualization again, but this time using QGIS.

Completing this chapter, you will learn about the following topics:

- The basics of GIS and vector data
- Coordinate transformation in R and QGIS
- Visualizing quantitative and qualitative data in maps in R
- Using **OpenStreetMap (OSM)** as background

GIS in R

GIS is a combination of software and data that informs us about the location of something and its relation to others. In GIS, every dataset is associated with a coordinate system, which is a system for representing the locations of different geographic features and different measurements. There are two main types of coordinate systems: **geographic coordinate systems (GCS)** and **projected coordination systems**. One example of GCS is using latitude-longitude, and one example of a projected coordination system is the transverse Mercator system. Whereas GCS uses a three-dimensional spherical surface, the projected coordination system uses two dimensions for representing spatial data. Data is used in GCS to define the position of the spheroid in relation to the center of the earth; a very commonly used GCS is **WGS 84**.

Data types in GIS

There are two main data types in GIS:

- **Vector** data
- **Raster** data

Vector data

Vector data is good for representing categorical and multivariate data. It also has attribute tables where every record or row corresponds to a feature or an object and every column corresponds to different attributes. Vector data has three different types: **points**, **lines**, and **polygons**. Points data normally refers to data collected at some point in space. Points connect to each other to become lines and represent features such as highways, paths of anything, and so on. If those lines add up to make a closed shape, we get a polygon.

One commonly used vector data format is shapefile, which is mainly specific to ArcGIS but is widely used now. Shapefile is used to describe vector features such as points, lines, and polygons. For doing so, shapefile uses three to seven files for a single map file to describe different components such as geometries or shape format, projection, attributes, and so on. In shapefile, a file with an `.shp` extension represents geometry, an `.shx` file represents the positional index of the feature geometry, and a file with a `.dbf` extension represents attribute formats in the dBASE IV format. There are also other files such as `.prj` for projection, `.ixs` for geocoding indexes, and some more. But `.shp`, `.shx`, and `.dbf` are essential files for constituting a shapefile for representing vector data.

Raster data

Raster data is used for continuous data and stores data in a grid-like arrangement. Raster is good for measuring continuous features such as elevation, vegetation, and so on.

Now, in the next section, we will have a look at how to work with different vector data in R, starting with point data.

Plotting point data

Location is a point characterized by its coordinates. Point data comprises attributes of a location or data collected on a parameter from different points. Shapefile is a popular data format for vector data. In R, the sp package loads shapefile as SpatialPoints and if it contains attributes, it is saved as SpatialPointsDataFrame. We can import a shapefile into R, using the readOGR() function of the sp package. We have to provide the path to the shapefile as the first parameter for the readOGR() function and the name of the shapefile is the second parameter. Here, we have to provide the name of the shapefile without the .shp extension. So, if we want to import a shapefile containing points, we can do so by writing the following:

```
# SpatialPoints
library(sp)
library(rgdal)
library(maptools)
map = readOGR("F:/Hands-on-Geospatial-Analysis-Using-R-and-QGIS/Chapter
02/Data","indicator")
plot(map)
```

We see the following map, which just shows point data as a plus point indicating different districts in Bangladesh:

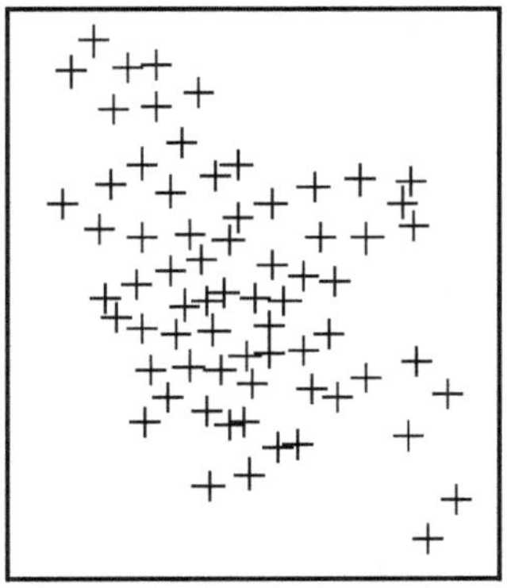

Now, let's check the class of this data `map`:

```
class(map)
```

The class is `SpatialPointsDataFrame`.

Importing point data from Excel

We have many arbitrary measures collected for different districts of Bangladesh along with their point coordinates; these measures contain both numeric and categorical values.

Now, we will import this data into R using `read.csv()` as follows:

```
bd_val = read.csv("F:/Hands-on-Geospatial-Analysis-Using-R-and-
QGIS/Chapter02/Data/r_val.csv", stringsAsFactors = FALSE)
```

We check the structure of this dataset `bd_val` using `str()`:

```
str(bd_val)
```

We get the following output:

```
'data.frame':   66 obs. of  6 variables:
 $ X    : int  1 2 3 4 5 6 7 8 9 10 ...
 $ lat  : num  22.9 22.6 22.5 22.4 22.1 ...
 $ lon  : num  90.2 90.2 90 90.7 90.2 ...
 $ names: chr  "Barisal" "Jhalakati" "Pirojpur" "Bhola" ...
 $ value: int  23 43 56 34 54 87 19 87 45 73 ...
 $ ind  : chr  "A" "B" "C" "D" ...
```

We see that the type of `bd_val` is dataframe. Now, we convert this into `SpatialPointsDataFrame` by using `coordinates()` and by specifying which columns contain the longitude and latitude of these points.

```
# Convert it into SpatialPointsDataframe
coordinates(bd_val) = c("lon", "lat")
str(bd_val)
```

```
Formal class 'SpatialPointsDataFrame' [package "sp"] with 5 slots
  ..@ data       :'data.frame': 66 obs. of  4 variables:
  .. ..$ X     : int [1:66] 1 2 3 4 5 6 7 8 9 10 ...
  .. ..$ names: chr [1:66] "Barisal" "Jhalakati" "Pirojpur" "Bhola" ...
  .. ..$ value: int [1:66] 23 43 56 34 54 87 19 87 45 73 ...
  .. ..$ ind  : Factor w/ 5 levels "A","B","C","D",..: 1 2 3 4 5 1 2 3 4 5
  ..@ coords.nrs : int [1:2] 3 2
  ..@ coords     : num [1:66, 1:2] 90.2 90.2 90 90.7 90.2 ...
  .. ..- attr(*, "dimnames")=List of 2
  .. .. ..$ : NULL
  .. .. ..$ : chr [1:2] "lon" "lat"
  ..@ bbox       : num [1:2, 1:2] 88.3 21.5 92.4 26.3
  .. ..- attr(*, "dimnames")=List of 2
  .. .. ..$ : chr [1:2] "lon" "lat"
  .. .. ..$ : chr [1:2] "min" "max"
  ..@ proj4string:Formal class 'CRS' [package "sp"] with 1 slot
  .. .. ..@ projargs: chr NA
```

Now, plot this using `plot()`:

```
plot(bd_val, col = "blue", pch = 19)
```

Now, we get the following map with blue dot for each point:

Plotting lines and polygons data in R

Lines data consists of lines and, in the `sp` package of R, it is stored as
a `SpatialLines` class. If it contains attributes, it is saved as `SpatialLinesDataFrames`.
Similarly for polygons, without attributes the class is defined as `SpatialPolygons` and
with attributes the class is defined as `SpatialPolygonsDataFrame`.

We will load a shapefile consisting of lines attributes, which will be treated as
a `SpatialLinesDataFrames` class in R. Let's load the shapefile of highways in Dhaka,
Bangladesh:

```
# SpatialLines
highway = readOGR("F:/Hands-on Geospatial Analysis Using R and
```

```
QGIS/Chapter02/Data","dhaka_gazipur")
plot(highway)
```

This gives us the following map:

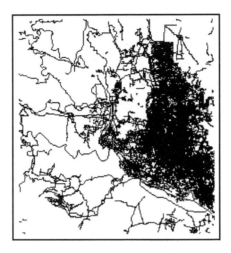

Now, we will read polygons (a map of Dhaka saved as dhaka.shp) into R as SpatialPolygonsDataFrame, and will plot this:

```
map_dhaka = readOGR("F:/Hands-on-Geospatial-Analysis-Using-R-and-
QGIS/Chapter02/Data","dhaka")
plot(map_dhaka)
```

This gives the following gray map:

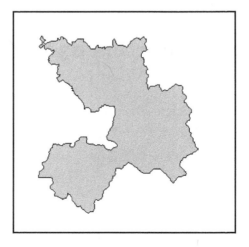

Now, let's have a look at the structure of this `SpatialPolygonsDataFrame` to understand how it is being stored and how to access and manipulate it:

```
# Use max.level = 2 to show a reduced or succinct structure
str(map_dhaka, max.level = 2)
```

The structure of `map_dhaka` contains:

- @ `data`: This contains all the attribute information or it contains data.
- @ `polygon`: This stores information on polygons or coordinates.
- @ `bbox`: This contains information on the extent of the map or the coordinates of two corners of the bounding box.

These three parts of the structure are pointed out in the following screenshot:

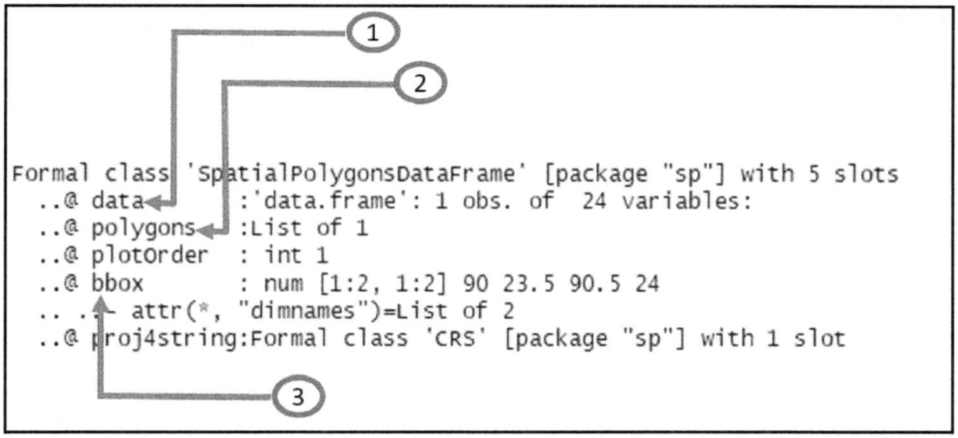

In the preceding screenshot, we can see that it has five slots and each of these can be accessed using @. If we want to access data and see the first five rows of it, we can do so by doing the following:

```
# load another map
map_bd = readOGR("F:/Hands-on-Geospatial-Analysis-Using-R-and
QGIS/Chapter02/Data", "BGD_adm3_data_re")
head(map_bd@data)
```

We can see the first five rows of the attribute table now:

```
  ID_0 ISO     NAME_0 ID_1 NAME_1 ID_2   NAME_2 ID_3    NAME_3                          VARNAME_3 NL_NAME_3  HASC_3 TYPE_3
0   23 BGD Bangladesh 325 Barisal 3095   Barisal 3624   Barisal                              <NA>     <NA> BD.BA.BS   Zila
1   23 BGD Bangladesh 325 Barisal 3095   Barisal 3625 Jhalakati Jhalakati|Jhalakhati|Jhalkathi     <NA>     <NA> BD.BA.JK   Zila
2   23 BGD Bangladesh 325 Barisal 3095   Barisal 3626  Pirojpur                              <NA>     <NA> BD.BA.PR   Zila
3   23 BGD Bangladesh 325 Barisal 3096 Patuakhali 3627     Bhola                              <NA>     <NA> BD.BA.BL   Zila
4   23 BGD Bangladesh 325 Barisal 3096 Patuakhali 3628   Borgona                  Barguna|Borgona     <NA> BD.BA.BG   Zila
5   23 BGD Bangladesh 325 Barisal 3096 Patuakhali 3629 Patuakhali                            <NA>     <NA> BD.BA.PT   Zila
  ENGTYPE_3 VALIDFR_3 VALIDTO_3 REMARKS_3 Shape_Leng Shape_Area Value value1 value2           value3
0  District   Unknown   Present      <NA>   9.978729 0.19656914    90   1234     67            Agree
1  District   Unknown   Present      <NA>   2.677218 0.06195496    NA   3456     45  Strongly Agree
2  District   Unknown   Present      <NA>   3.534049 0.10831397    NA   1349     65        Not Sure
3  District   Unknown   Present      <NA>   8.704750 0.15616449    NA   1200     22         Disagree
4  District   Unknown   Present      <NA>   5.713298 0.11849395    NA   2100     45 Strongly Disagree
5  District      1946   Present      <NA>  13.467342 0.21025806    NA   2090     65            Agree
```

Now, let's examine @ polygons:

```
str(map_bd@polygons, max.level = 2)
```

What we find is another list of 66 where each list is again 66 polygons and each again has five slots just as map_bd has. The following is a snapshot of the first few lines of output. The remaining lines have not been shown here for the purposes of keeping the example succinct:

```
List of 66
 $ :Formal class 'Polygons' [package "sp"] with 5 slots
 $ :Formal class 'Polygons' [package "sp"] with 5 slots
 $ :Formal class 'Polygons' [package "sp"] with 5 slots
 $ :Formal class 'Polygons' [package "sp"] with 5 slots
 $ :Formal class 'Polygons' [package "sp"] with 5 slots
 $ :Formal class 'Polygons' [package "sp"] with 5 slots
 $ :Formal class 'Polygons' [package "sp"] with 5 slots
 $ :Formal class 'Polygons' [package "sp"] with 5 slots
 $ :Formal class 'Polygons' [package "sp"] with 5 slots
 $ :Formal class 'Polygons' [package "sp"] with 5 slots
 $ :Formal class 'Polygons' [package "sp"] with 5 slots
 $ :Formal class 'Polygons' [package "sp"] with 5 slots
 $ :Formal class 'Polygons' [package "sp"] with 5 slots
 $ :Formal class 'Polygons' [package "sp"] with 5 slots
 $ :Formal class 'Polygons' [package "sp"] with 5 slots
```

That means we can also access these lists' slots using @ and any of the five slots previously discussed. We now access the 6th element of map_bd and investigate its structure as follows:

```
# 6th element in the Polygons slot of map_bd
sixth_element = map_bd@polygons[[6]]
# make it succinct with max.level = 2 in str() for the 6th element of the
bd@Polygons
str(sixth_element, max.level = 2)
```

We can see the structure of the `6th element` of the polygon now:

```
Formal class 'Polygons' [package "sp"] with 5 slots
  ..@ Polygons :List of 83
  ..@ plotorder: int [1:83] 82 74 41 77 79 51 60 65 43 38 ...
  ..@ labpt    : num [1:2] 90.5 22.3
  ..@ ID       : chr "5"
  ..@ area     : num 0.21
```

Now, again check the structure of the 2nd polygon inside `sixth_element`. We can do so by writing the following:

```
# Structure of the 2nd polygon inside seventh_element
str(sixth_element@Polygons[[2]], max.level = 2)
```

```
Formal class 'Polygon' [package "sp"] with 5 slots
  ..@ labpt   : num [1:2] 90.3 21.8
  ..@ area    : num 9.08e-06
  ..@ hole    : logi FALSE
  ..@ ringDir: int 1
  ..@ coords  : num [1:27, 1:2] 90.3 90.3 90.3 90.3 90.3 ...
```

Now, we can access these slots and. for demonstration purposes only, we will access `coords` and then will plot it:

```
# plot() the coords slot of the 2nd element of the Polygons slot.
plot(sixth_element@Polygons[[2]]@coords)
```

This gives the following graph:

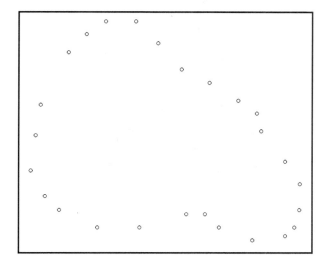

To access data elements of a `SpatialPolygonsdataFrame`, we can use either `$` or `[[]]` as we can do with a data frame. To access the column or attribute `NAME_3`, we can do the following:

```
map_bd$NAME_3
```

This will print all the values of the attribute `NAME_3`.

We can do the same using `[[]]` in the following way:

```
map_bd[["NAME_3"]]
```

Adding point data on polygon data

Now, we will plot point data on polygon data. For doing so, we will first plot `SpatialPolygonsDataFrame` (containing polygons) and then will add `SpatialPointsDataFrame` (containing points) using `points()`. We do so in the following way:

```
# Adding point data on polygon data
plot(map_bd)
points(bd_val, pch=19, col="blue")
```

This gives us the following map where points are overlaid on the polygon:

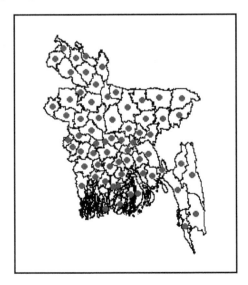

Changing projection system

To change projection, we can use `spTransform()` from the `rgdal` package. We can do so using a `CRS()` argument inside `spTransform()`:

```
map_bd = spTransform(map_bd, CRS("+proj=longlat +datum=WGS84"))
```

The preceding code sets the projection system to the longitude and latitude and the GCS to `WGS84`. If we want to change the projection to any other layer's (shapefile's) projection inside `CRS()`, we can write `proj4string()` to get the CRS of a new layer and then set it to that. For example, if we want to set the projection system of a layer `a` to the projection system of layer `b`, we can do so simply by writing the following:

```
a = spTransform(a, CRS(proj4string(b)))
```

Plotting quantitative and qualitative data on a map

We can plot quantitative values using the `choropleth()` of the `GISTools` package. We can generate a choropleth using the following commands:

```
# plot quantitative data
library(GISTools)
choropleth(map_bd, as.numeric(map_bd$value2))
```

We can also write a title and design this map further, but we will do these things in upcoming chapters. This gives us a nice little map:

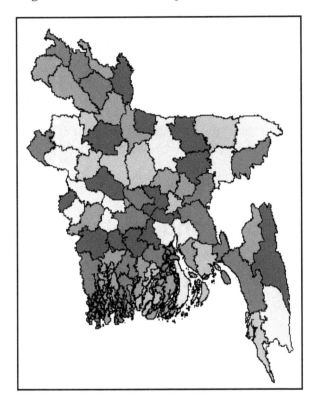

Using `spplot()`, we can also plot qualitative data. First, we need to convert this qualitative attribute or column of `SpatialPolygonsDataFrame` to a factor variable and use a suitable color range. We have a shapefile of Dhaka's divisions, which consist of seven districts each whose name is stored in the NAME_3 column, and our goal is to color different districts of Dhaka's divisions. Here we have picked seven colors from the `RColorBrewer` package as there are seven unique values for the NAME_3 column. Plotting to qualitative data (here the names of the districts) or coloring different districts can be done in the following way:

```
# Plot qualitative data
#install.packages("RColorBrewer")
library(RColorBrewer)
dhaka_div = readOGR("F:/Hands-on-Geospatial-Analysis-Using-R-and-
QGIS/Chapter02/Data","dhaka_div")
# check how many unique elements map_bd$NAME_3 has by writing
unique(dhaka_div$NAME_3)
```

```
unique(dhaka_div$NAME_3)
# There are 7 unique districts and so pick 7 colors
colors = colorRampPalette(brewer.pal(12, "Set3"))(7)
dhaka_div$NAME_3 = as.factor(as.character(dhaka_div$NAME_3))
spplot(dhaka_div, "NAME_3", main = "Coloring different districts of Dhaka
division", col.regions = colors, col = "white")
```

This gives us the following map with every polygon colored according to its name:

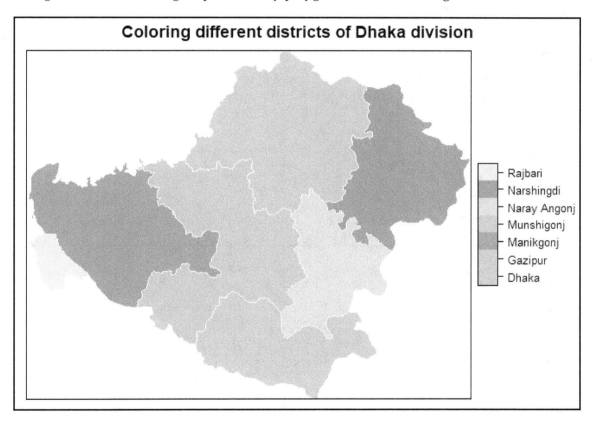

We will be learning easier ways to plot now in the next section.

Using tmap for easier plotting

We can map quantitative and qualitative values in a much easier way using the `tmap` package. For a simple choropleth map, we can just use the `qtm()` function and inside it, there are two arguments. The first one is the shapefile and the second one is the numeric column we want to use for making a choropleth map. First, we need to install and load it before we proceed further:

```
# Using tmap
install.packages("tmap")
library(tmap)
```

We will be using `BGD_adm3_data_re.shp` and will use the `value1` column for mapping. Let's load this dataset and check the structure of the attribute table associated with this shapefile:

```
# load the map
map_bd = readOGR("F:/Hands-on-Geospatial-Analysis-Using-R-and-
QGIS/Chapter02/Data","BGD_adm3_data_re")
#head(map_bd@data)
str(map_bd@data)
```

This shows that the numeric column `value1` is stored as a factor variable by R:

```
  str(map_bd@data)
  data.frame':   66 obs. of  23 variables:
  $ ID_0       : int  23 23 23 23 23 23 23 23 23 23 ...
  $ ISO        : Factor w/ 1 level "BGD": 1 1 1 1 1 1 1 1 1 1 ...
  $ NAME_0     : Factor w/ 1 level "Bangladesh": 1 1 1 1 1 1 1 1 1 1 ...
  $ ID_1       : int  325 325 325 325 325 325 326 326 326 326 ...
  $ NAME_1     : Factor w/ 6 levels "Barisal","Chittagong",..: 1 1 1 1 1 1 2 2 2 2 ...
  $ ID_2       : int  3095 3095 3095 3096 3096 3096 3097 3098 3098 3099 ...
  $ NAME_2     : Factor w/ 23 levels "Bandarban","Barisal",..: 2 2 2 18 18 18 1 4 4 5 ...
  $ ID_3       : int  3624 3625 3626 3627 3628 3629 3630 3631 3632 3633 ...
  $ NAME_3     : Factor w/ 64 levels "Bagerhat","Bandarbon",..: 3 24 53 4 6 52 2 9 12 7 ...
  $ VARNAME_3  : Factor w/ 30 levels "Bandarban|Bandarbon",..: NA 9 NA NA 2 NA 1 3 NA NA ...
  $ NL_NAME_3  : Factor w/ 0 levels: NA NA NA NA NA NA NA NA NA NA ...
  $ HASC_3     : Factor w/ 64 levels "BD.BA.BG","BD.BA.BL",..: 3 4 5 2 1 6 8 12 9 7 ...
  $ TYPE_3     : Factor w/ 1 level "Zila": 1 1 1 1 1 1 1 1 1 1 ...
  $ ENGTYPE_3  : Factor w/ 1 level "District": 1 1 1 1 1 1 1 1 1 1 ...
  $ VALIDFR_3  : Factor w/ 5 levels "1946","1974",..: 5 5 5 5 5 1 3 5 5 5 ...
  $ VALIDTO_3  : Factor w/ 1 level "Present": 1 1 1 1 1 1 1 1 1 1 ...
  $ REMARKS_3  : Factor w/ 0 levels: NA NA NA NA NA NA NA NA NA NA ...
  $ Shape_Leng : num  9.98 2.68 3.53 8.7 5.71 ...
  $ Shape_Area : num  0.197 0.062 0.108 0.156 0.118 ...
  $ Value      : num  90 NA NA NA NA NA 73 40 NA NA ...
  $ value1     : Factor w/ 60 levels "1020","1200",..: 3 39 4 2 12 11 16 32 41 49 ...
  $ value2     : Factor w/ 38 levels "17","19","21",..: 23 15 22 4 15 22 5 27 15 24 ...
  $ value3     : Factor w/ 5 levels "Agree","Disagree",..: 1 4 3 2 5 1 2 5 5 2 ...
```

Thus, we need to change its type to numeric. We do this in the following way:

```
map_bd$value1 = as.numeric(map_bd$value1)
```

Now, we are ready to use the `qtm()` function and, providing two arguments, the `SpatialPolygonsDataFrame` (the imported shapefile) and the column to be used:

```
qtm(shp = map_bd, fill = "value1")
```

The resulting map is as follows:

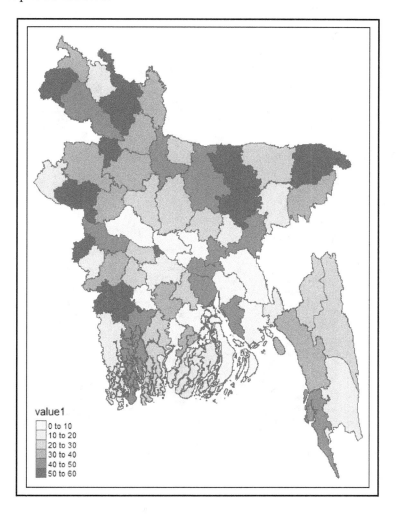

Similar to ggplot2, we can add subsequent layers in tmap for plotting. First, inside tm_shape(), we provide the SpatialPolygonsDataFrame, then we can add layers for borders, for fill, and also for style, among other features. Let's make a choropleth map again but this time using tmap() and also using multiple layers:

```
tm_shape(map_bd) +
  tm_borders() +
  tm_fill(col="value1") +
  tm_compass() + # This puts a compass on the bottom left of the map
  tmap_style("cobalt")
```

This gives us the following map:

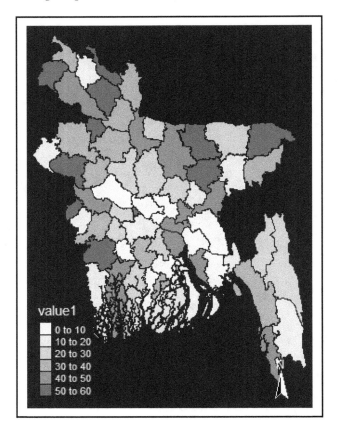

We can also use bubble size to correspond to the magnitude of the values using `tm_bubbles()` instead of `tm_fill()`.

```
tm_shape(map_bd) +
  tm_bubbles(size = "value1", style = "quantile") +
  tm_borders(col="orange3") # Add a colorful border
```

Now, we will see a beautiful map of Bangladesh with the values plotted as bubbles.

We can also label this map very easily with `tmap`. We need to use `tm_text()` as a new layer and specify which column it should use for labeling. As `map_bd` corresponds to the districts of Bangladesh and its name is represented by the `NAME_3` column, we will use that for labeling, using `text = "NAME_3"` inside `tm_text()` as follows:

```
# labeling
tm_shape(map_bd) +
  tm_fill(col = "value1", style = "quantile") +
  tm_borders() +
  tm_text(text = "NAME_3", size = 0.5)
```

We can also use `tm_credits()` for adding a credit to the map or to just put some relevant text on the map. It has a position argument, and by using this we can position it in different places on the map. Now, we are going to add credits and some text regarding the quantitative variable we are using as follows:

```
tm_shape(map_bd) +
  tm_fill(col = "value1", style = "quantile", title = "Value of quantitative
indicator", palette = "Blues") +
  tm_borders(col = "grey30", lwd = 0.6) +
  tm_text(text = "NAME_3", size = 0.5) +
  tm_credits("Source: Author", position = c("right", "top"))
```

Vector data in QGIS

Now, we will work with a shapefile in QGIS and find out its projection system. First, we import a shapefile named `BGD_adm4.shp` from the folder `Data` under `Chapter 2`. To recall how to import a vector file, refer back to `Chapter 1`, *Setting Up R and QGIS Environments for Geospatial Tasks*.

Now, we will see the shapefile as follows:

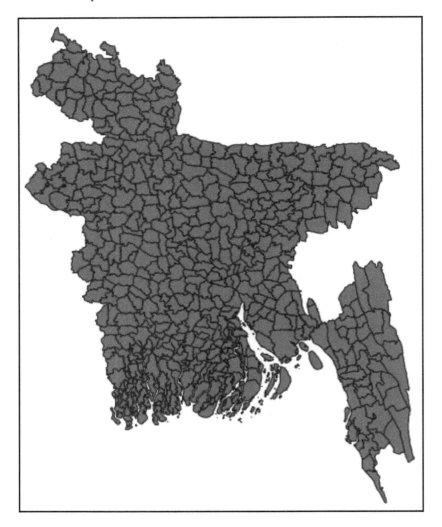

Now, to check out the projection system it is using, we need to click on **Project Properties** under **Project** in the menu bar as instructed in the following screenshot:

1. Click on **Project** and then **Properties...** under it:

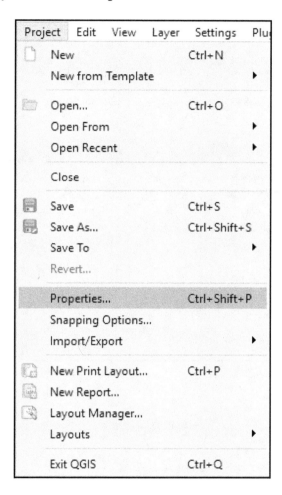

2. Now, click on **CRS**:

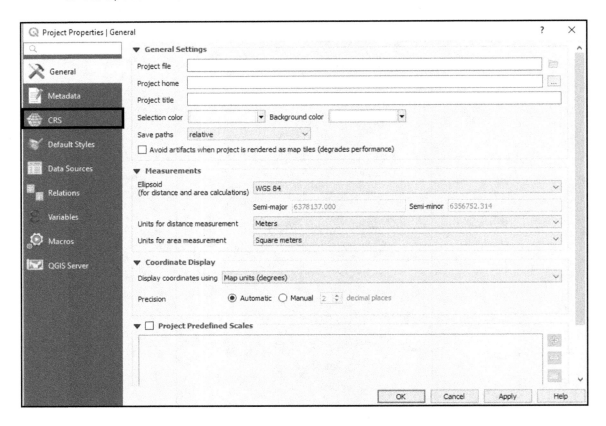

Now, we can see the coordinate system (**a**) and the spatial extent and projection (**b**) as shown in this screenshot:

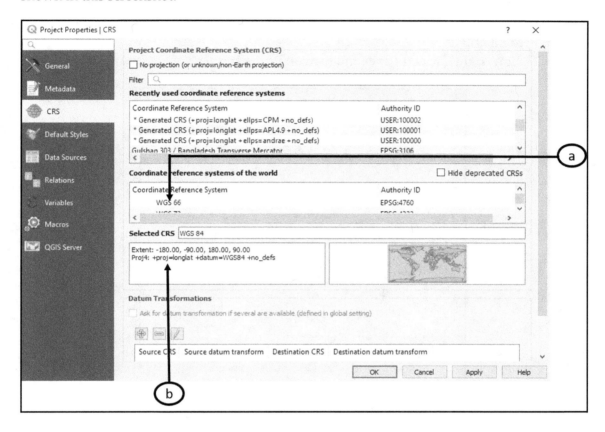

We can also select and change CRS from the **Layer** panel. We can do so using the following steps:

1. Right-click on the `BGD_adm4` layer in the **Layer** panel. Click on **Properties** as highlighted.
2. Now, right-click on **Layer** and then on **Properties**:

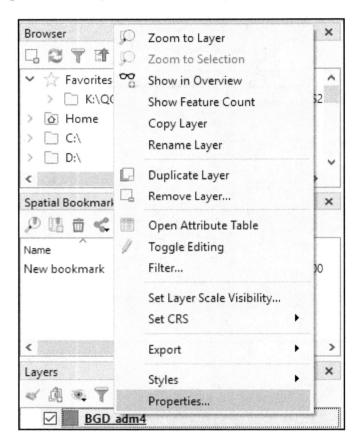

3. Now, left-click **Source**.

4. We can see CRS now. We can also change CRS here by selecting our preferred CRS and then by pressing **Apply** and subsequently by pressing **OK**:

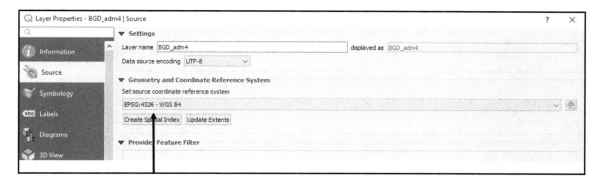

For simplicity, we will work now with `BGD_adm3.shp` and there is no snapshot of this as we know how to load a vector file by now. In its attribute table, we can find its unique identifier field `ID_3`.

Adding Excel data in QGIS using joins

Now, we will add data from an Excel file called `value.xlsx` to this attribute table (this file is located in the `Data` folder). This file contains information from a fictitious survey and has four columns: the first one `ID_Dis` (a unique identifier matching with the shapefile's attribute table), two columns with quantitative values, and one column with a qualitative value showing the opinion of participants regarding an imagined view. We can add this Excel file in the same way as we have done with vector (shapefile) data. Then, we can add this Excel file following the steps outlined next:

1. First, load `BGD_adm3.shp` and then right-click on `BGD_adm3` and then click **Properties**.

2. Now, left-click **Joins** and then click on the plus sign situated in the lower left of the window:

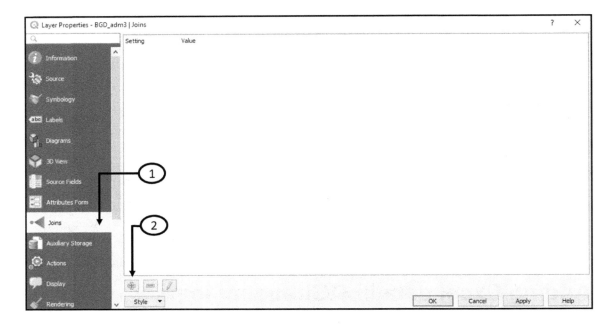

3. We will now see our Excel file `value Sheet1` in the **Join layer**. We can select the unique ID of the Excel file by selecting `ID_Dis` from the drop-down for the **Join field**. Now, we will select the unique ID of the shapefile, `ID_3`, for the **Target field**. After that, click **OK**:

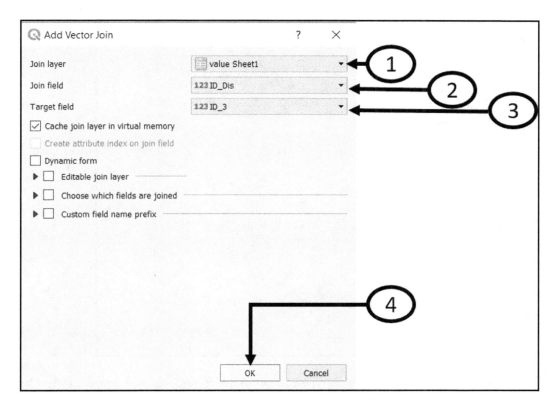

4. Finally, we click **Apply** and then click **OK**.

Now, we should have the values from our Excel file added to the attribute table of the shapefile. We can check whether that's the case by investigating the attribute table as follows:

1. Open the attribute table of `BGD_adm3`.
2. Three new columns have been added to the attribute table. Note that the names now look different than the original ones.

Next, we will save this shapefile with the name `BGD_adm3_val` using the following steps:

1. First, right-click on the `BGD_adm3` layer and then press **Save as** as highlighted:

2. Select ESRI Shapefile for the file format. After that, click on a box indicated by ... under **File name** to browse to the file.
3. Then, browse to a folder where we want to save it and then click **Save**.
4. Left-click **OK** and a new shapefile named `BGD_adm3_data` with columns from the Excel file is saved.

Adding CSV layers in QGIS

Suppose we have the value of an arbitrary indicator collected from all the districts (administrative units) of Bangladesh. This result is saved in a CSV file, named `indicator.csv`, and it contains the longitude and latitude values of all the locations as *x* and *y* columns in the file, and it has another column containing the values of that indicator. Now, we want to plot it on a map or on `BGD_adm3_data.shp` and customize the visualization. So, first we open the shapefile that we created in the preceding segment and we can then follow the steps outlined hereafter to map the point data on the map or shapefile:

1. Click **Layer** in the menu bar. Then, click **Add Delimited Text Layer** under **Add Layer**:

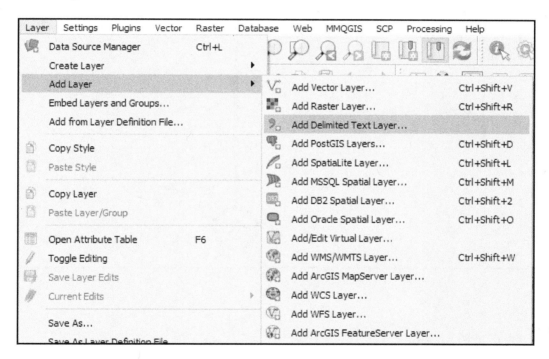

2. Now, browse to the folder where the `indicator.csv` file is located (under the `Data` folder for our case). Select the CSV file **indicator.csv** and left-click **Open**.
3. Now, click on **Geometry Definition** to set the coordinate system.

4. Now, select x for the **X field** (this sets the longitude), select y for the **Y field** (setting latitude). Select WGS 84 for **Geometry CRS**, then press **Add**, and then press **Close**.

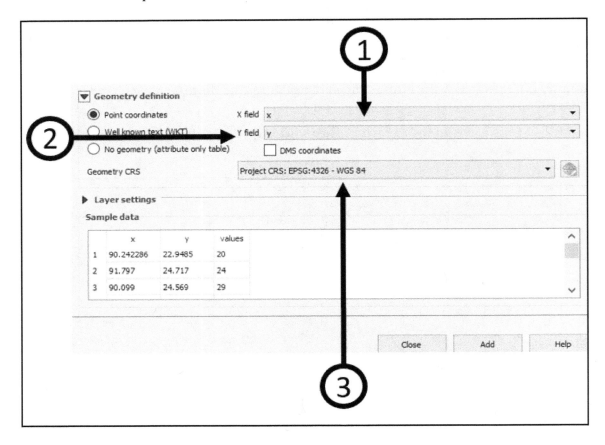

Now, we can see that values from the CSV file are plotted on the map. Note how the values from the CSV file are plotted according to longitude and latitude. It's worth noting that the indicator layer should come after the `BGD_adm3_data` layer if we want to map those points on the top of the shapefile. Our map should look such as the following:

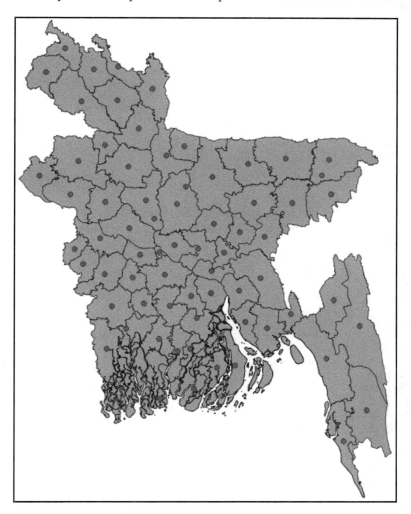

We can select different symbols for these points mapped on the shapefile. For example, we can select a green-colored pentagon shape from QGIS to be the symbol for these values. We can do so following these guidelines:

1. Double click the **indicator** layer:

2. Now click on **Symbology**, then select **Single symbol** from the top drop-down menu. Then select a **Color**, select a **Symbol**, then click **Apply**, and then click **OK**:

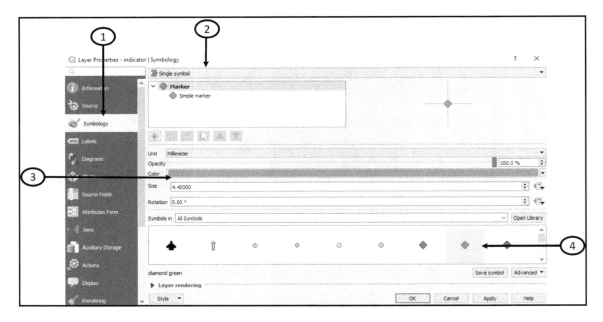

3. Now, we will find a plot according to our specification; that is, we can see the symbols and also will observe that the colors of the symbols have changed.

We have just seen one example of changing symbols, but as we can see, there are numerous other options for selecting different symbols and colors. It is encouraged that the readers explore some of the options under **Symbology** to get a better grasp of this.

Now, we will work with the attribute table of the shapefile `BGD_adm3_data` itself. We will work with qualitative data that is in the column named `value Sh_2`, which contains values such as **Agree, Disagree**, and so on. Note that this column name could be different on your computer and if you have followed sequentially along the steps outlined so far, this column should be the last of the attribute table for this shapefile. Now, we will learn to stylize vector data (shapefile) according to qualitative variables, as outlined as follows:

1. First double-click on the `BGD_adm3_data` layer and then select **Symbology** in the **Layer Properties** and select:
    - **Categorized**
    - Under **Columns** select **value Sh_2**
    - Then, left-click **Classify**:

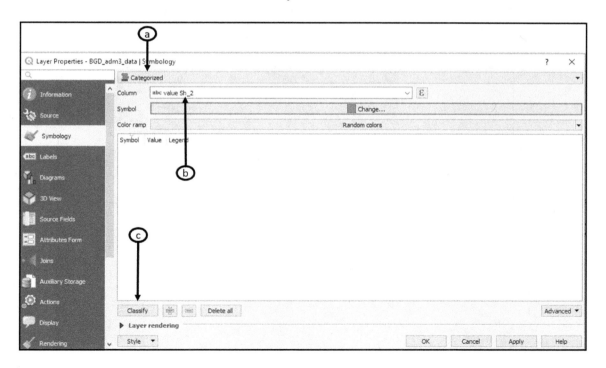

2. Now, we will find that the map is filled with different colors corresponding to the different values of the `value Sh_2` column:

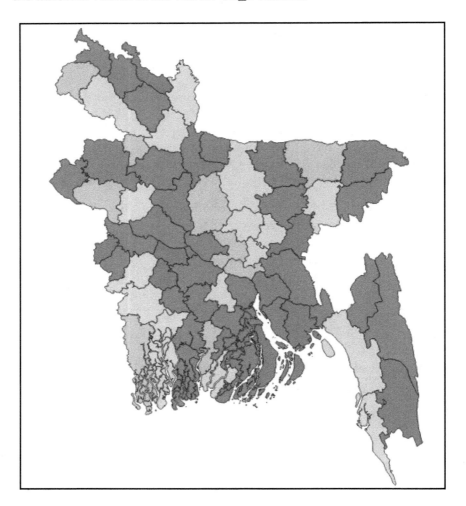

We can also label this shapefile according to the value of categorical attribute. For the preceding case, we need to click **Labels** of **Layer Properties** and take the following steps:

1. Open the BGD_adm3_data layer.
2. Select **Single labels**, then select value Sh_2 for the **Label with** field. Then, click **Apply** and then **OK**:

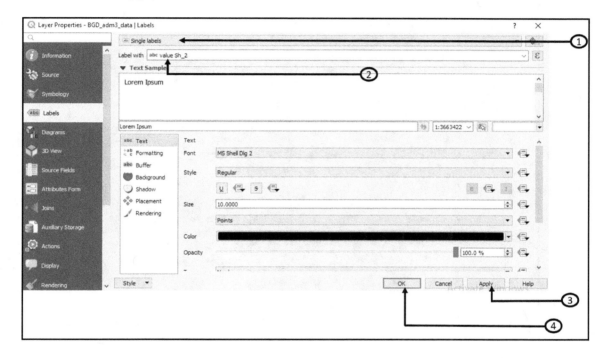

3. We get a labeled map now:

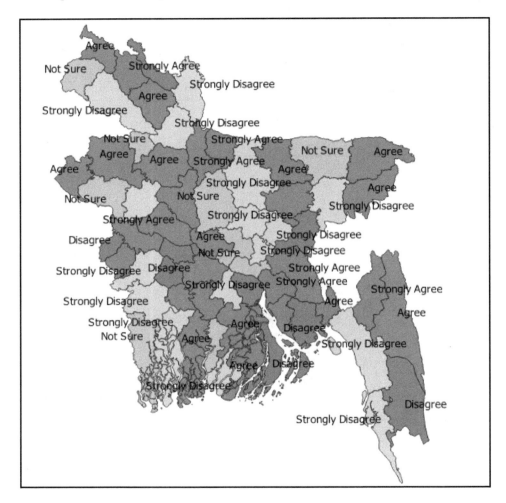

Now, using **Graduated** from QGIS, we can classify and map numeric variables. We will use the Jenkins method which classifies data so it has the most homogeneity within classes. The steps are shown in the following screenshot:

1. Double-click on the layer `BGD_adm3_data`.
2. Select **Graduated**, then select **Column** `value Sh_1`. Select **Natural Breaks (Jenks)** for the method of classification and then click **Classify**.

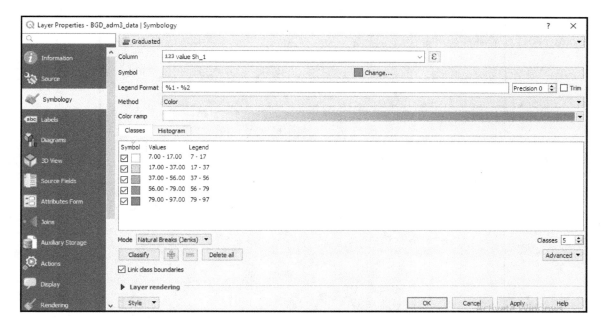

3. Now, click on **Color ramp** and then hover the mouse over all the color ramps and select any color ramp (for this example, we have stuck to the default). Now, click on **Apply** and then on **OK** to get the map:

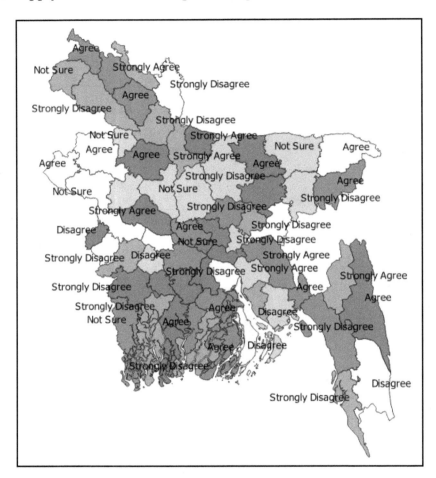

Showing multiple labels using text chart diagrams

Text chart diagrams show attribute values in a circle by dividing the circle into horizontal slices where each slice shows the value of each attribute we intend to show. It is useful for showing multiple values on the map at the same time.

Using the **Diagrams** tab from **Layer Properties**, we can label our map with multiple attributes of both numerical and categorical types. Suppose we want to map the last three attributes of BGD_adm3_data in the map; we can do so using **Text diagram**. The steps are outlined as follows:

1. Open the BGD_adm3_data layer by double-clicking. Double-click this layer and select the **Labels** tab and click on it.

2. Now, left-click on the **Diagrams** tab and select **Text diagram** under **No diagrams**:

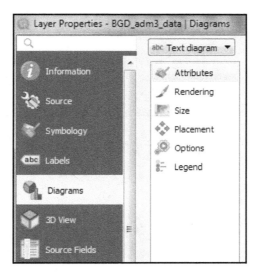

3. Select the attributes we want to show on the map. We select three attributes: **value Shee**, **value Sh1**, and **value Sh2** here. To do this, we first select **Attributes**, then we see a list of attributes under **Available attributes**, we select any of the attributes that we want to show on the map and then click on the green + sign to send it to the right panel under **Assigned attributes**. We do this for all three attributes and then we press **Apply** and then **OK**:

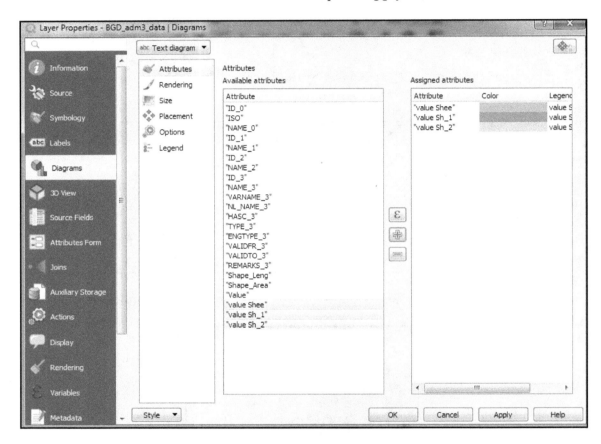

4. Now, we get a map labelled with all three attributes:

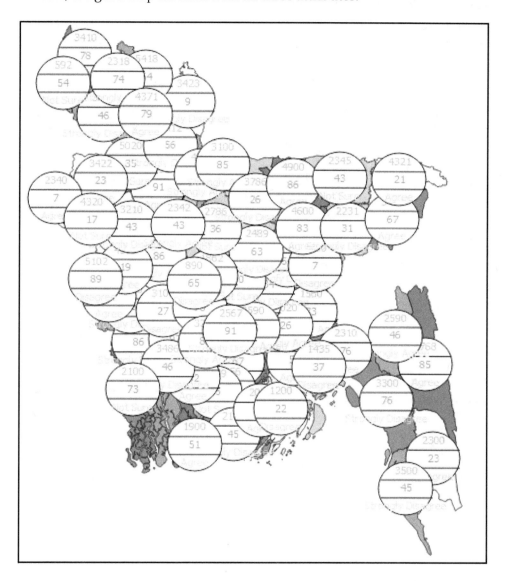

Adding a background map

We can add a background to our map of vector data by using the available plugins of QGIS. Here, we will add an OSM background by using the **QuickMapServices** plugin. The steps for doing this are outlined as follows:

1. Click on **Manage and Install Plugins...** under **Plugins**.
2. In the **Plugins** window, search for QuickMapServices and then click on **Upgrade plugin**. After the installation is completed, press **Close**:

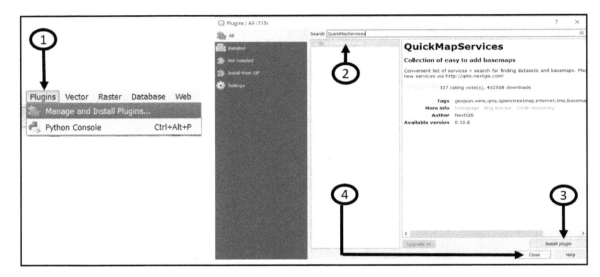

3. Add the shapefile `indicator_point.shp`. We need to enable the **on the fly** CRS and write `Pseudo` and select **Pseudo Mercator** for compatibility with OSM:

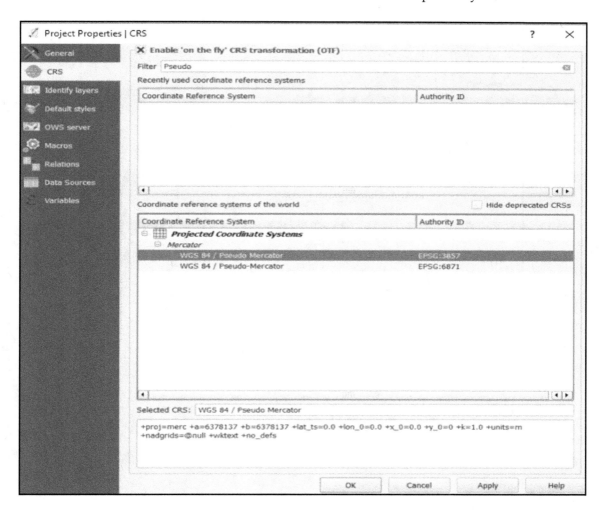

4. We can then add OSM by clicking the **Web** menu, followed
    by **QuickMapServices**, and then **OSM** and finally by clicking on **OSM Standard**:

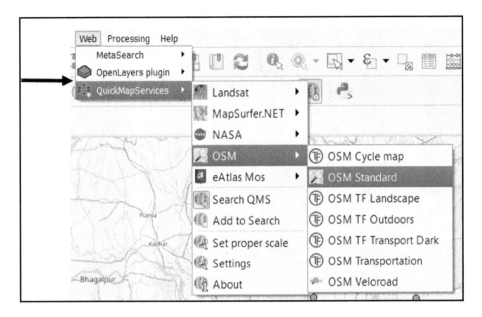

5. Ultimately, we will get the following map with OSM at the background:

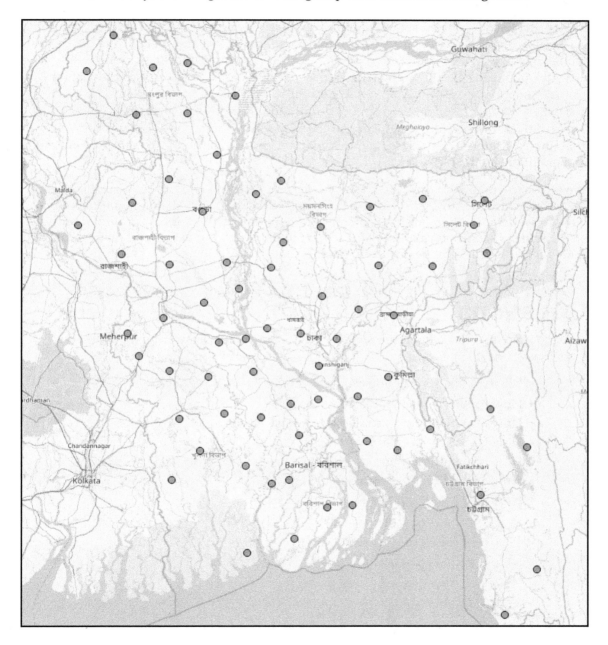

Summary

In this chapter, we learned about the basics of GIS and learned particularly how vector data is stored in R and QGIS. We learned to import point data from an Excel file and how to plot it on a map in R. We used the `ggmap` package of R to accomplish this task along with various options for plotting these. The `sp` package of R has six main classes for dealing with spatial data: `SpatialPoints`, `SpatialPointsDataFrame`, `SpatialLines`, `SpatialLinesDataFrame`, `SpatialPolygons`, and `SpatialPolygonsDataFrame`. We learned to use the `readOGR()` function of the `sp` package for importing shapefiles containing different vector classes. In this chapter, we also learned to visualize quantitative and qualitative data in R.

We also learned how to use QGIS for vector data importation and how to visualize those. Importing Excel files to QGIS was also discussed, along with visualizing quantitative and qualitative data. Further, adding text diagrams and adding OSM as background maps was also discussed. This naturally leads to the next topic, which is creating and editing data and that will be discussed in the next two chapters.

Questions

After completing this chapter, readers should be comfortable answering the following questions:

- What are the spatial data classes available in R (`sp` package)?
- How can users visualize spatial data in R?
- How can users import shapefiles and use background maps in R?
- How is CRS set in QGIS?
- How is quantitative and qualitative data visualized in QGIS?
- How are text diagrams used in QGIS?
- How may background maps be added?

Further reading

When going through vector data manipulation in R, we have just touched the surface of what could be possible using R. We will cover other data manipulation techniques as we step on through new chapters. But if you want to have a more in-depth discussion about these topics, you can have a look at the book *Learning R for Geospatial Analysis* by Michael Dorman and *An Introduction to R for Spatial Analysis and Mapping* by Brunsdon and Comber. For QGIS, if you want better coverage of visualizing and stylizing vector data, *QGIS by Example* by Bruy and Svidzinska and *Mastering QGIS* by Menke et al. are two great reads.

3
Creating Geospatial Data

In previous chapters, we learned how to use R and QGIS for basic geospatial tasks, such as loading vector data, visualizing it, and stylizing it. We've learned the basics of GIS and vector data, how R and QGIS store them, and more. This chapter will introduce you to the creation of data and the editing functionalities of spatial data offered by R and QGIS. We'll be introduced to different data sources and how to download data from one of these sites. Often, we need to create a digitized map from a printed map, and we'll be covering that in this chapter. After this, we'll learn how to create point, line, and polygon data using QGIS. We'll then learn how to add features to a shapefile and how to create and use a spatial database. These skills will enable us to work more efficiently with spatial data.

The following topics will be covered in this chapter:

- Getting data from the web
- Creating vector data
- Digitizing a map
- Working with databases

Getting data from the web

Working with geospatial data will often require us to use a spatial file that delineates country borders, roads, railways, rivers, coastlines, and so on. Sometimes, it's hard to manage all of this data by yourself. Luckily, we can download free shapefiles or vector data and raster from a number of websites for free. Here are three such websites:

- Natural Earth (`www.naturalearthdata.com/downloads/`)
- DIVA-GIS (`http://www.diva-gis.org/gdata`)
- EarthExplorer (`https://earthexplorer.usgs.gov/`)

Downloading data from Natural Earth

This website (`www.naturalearthdata.com/downloads/`) contains data on three different scales: large, medium, and small. The first two types of data contain shapefiles for cultural, physical, and raster data, whereas small-scale data is available for cultural and physical data only. The following is a screenshot of this page:

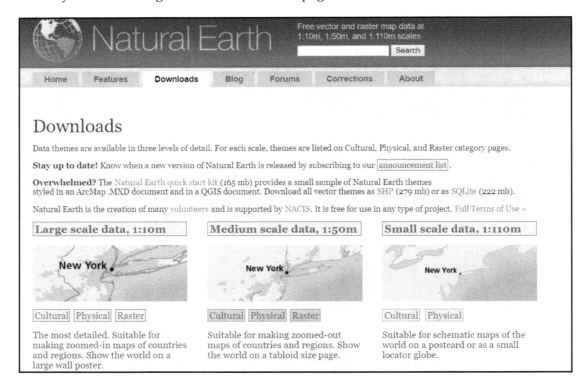

Now, if we click **Cultural**, a page will appear where we can download shapefiles for countries, subdivisions of countries, boundary lines, roads, railways, airports, urban areas, and more. For **Physical**, we can download coastlines, land, minor islands, reefs, oceans, rivers, and more. This data will be downloaded as a ZIP file and, after extracting it, we can use either R or QGIS to open and view the data.

Downloading data from DIVA-GIS

Using DIVA-GIS (`http://www.diva-gis.org/gdata`), we can download different types of shapefile and raster data easily. This website looks similar to this:

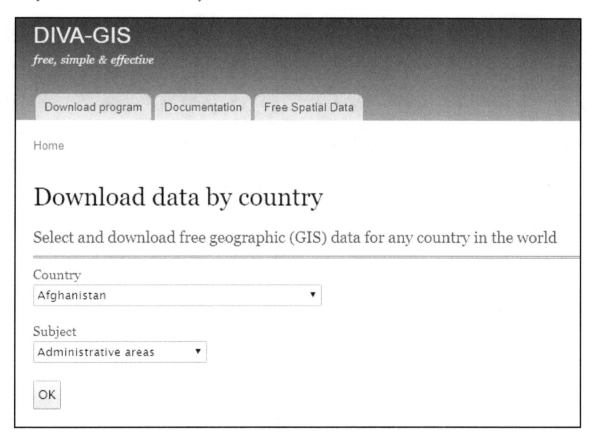

Again, after downloading the data, we can use R or QGIS for further processing of the data.

Downloading data from EarthExplorer

EarthExplorer allows us to download raster images of almost all parts of the world from multiple sources over a varying range of time periods. Navigating through this page is a little more complicated than for the previous two pages but, with a little practice, you can get the hang of it and download the images you need. If we want to download images from this file, we need to register with the website. The front page of this website looks like this:

The steps for downloading images from this website are as follows:

1. Write the name of the area in the search box. Here, we'll be downloading data for New York, and so we write `New York` in the box for **Address/Place**.

2. Click on **Show**:

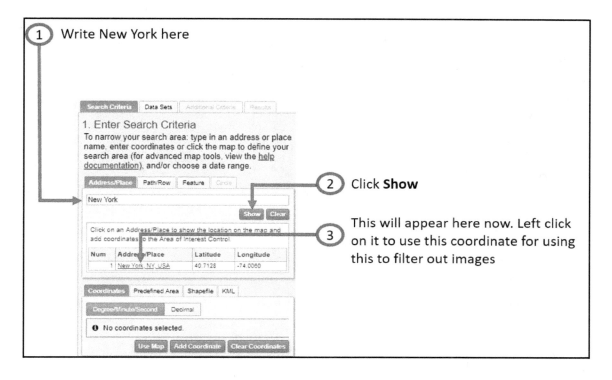

3. The coordinates of New York will be shown under **Degree/Minute/Second**. We'll need to define the **Data Sets** that we want to search for in this area:

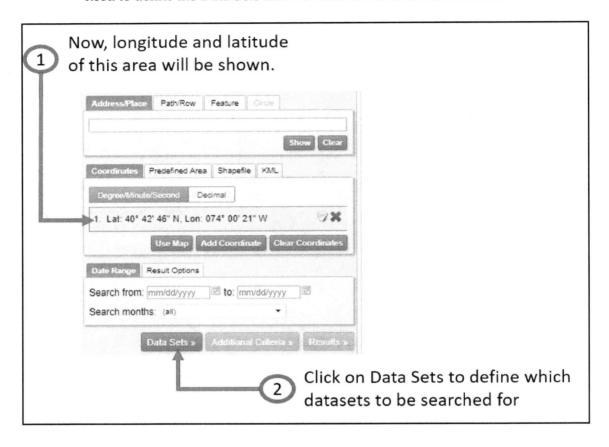

4. We need to select the data sources we want to look at. We can select one or multiple sources, and then images will be searched for according to these criteria. Suppose we select only Landsat 11 and Landsat 7 as sources. Were that the case, we would then need to click **Results** to see all of the available images:

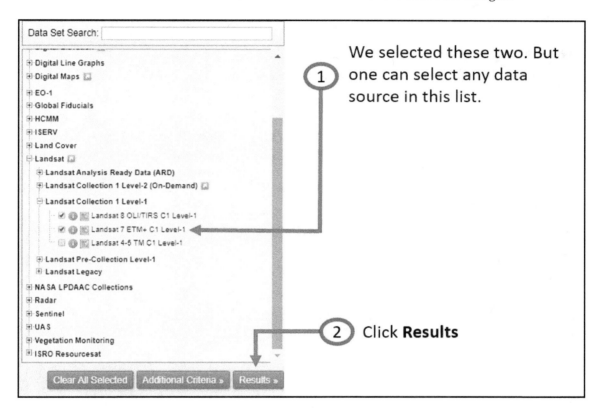

5. We'll see all of the images matching our criteria.
6. Click on the download link.
7. We'll then see a new window with matching images and download options. Click on the download icon to download this raster in GeoTIFF format.

Creating vector data

Often, we'll be required to create our own data by drawing point, line, or polygon data. In QGIS, we can do so very easily by specifying vector data when creating a shapefile.

Creating point data

Creating point vector data can be very easily accomplished in QGIS. We can create point data by following these steps:

1. We need to create a new shapefile by clicking on the **Layer** menu, clicking or hovering on **Create Layer**, and then clicking on **New Shapefile Layer**.

2. A new window will appear now. We browse to the folder where we want to save the shapefile and give it a name here it's named **points**. Select **Point** as **Geometry Type**. After that, we save it by clicking **OK**. Now we'll see that this layer is added under the **Layers** panel:

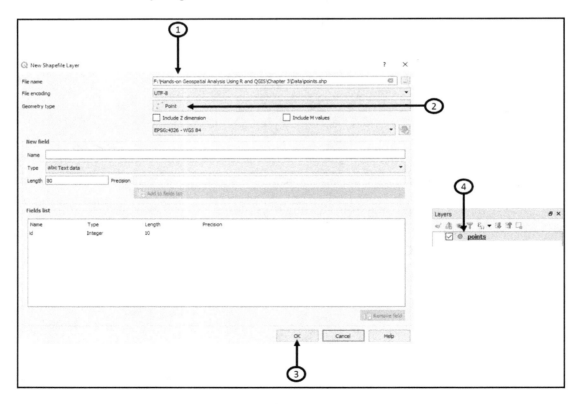

Creating polygon data

Creating line and vector data in QGIS is a very similar process to the workflows we followed for creating point vector data. Here, we'll look at some of the steps that we need to take in creating polygon data:

1. We need to create a new shapefile by clicking on **New Shapefile Layer** as shown for point data. Then, we need to select **Polygon** for **Geometry Type**. Later, we add a field to this layer by clicking inside the field for **Name** and specifying a name for the field; we've specified val, which is a numeric value, and so we've to select **Decimal number** under **Type** and then click on **Add to fields list**, as can be seen from the following screenshot:

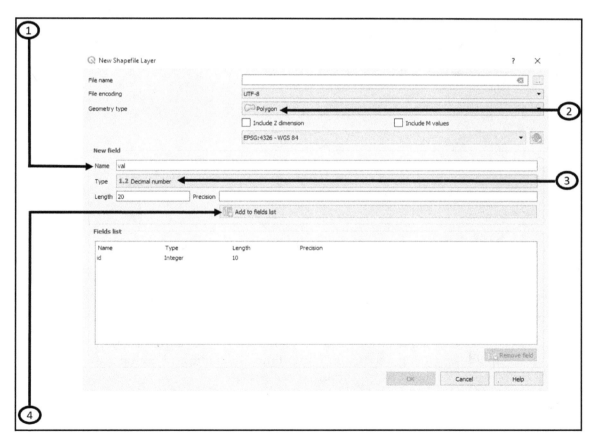

2. We can add multiple fields of different data types. Here, we'll also add another filed called **Name**, which is of the text type, by following the previous instructions. We create a field, Name, we specify its **Type** to be of Text data, and then we click on **Add to fields list**. We can follow similar steps for creating line data. These steps are shown here:

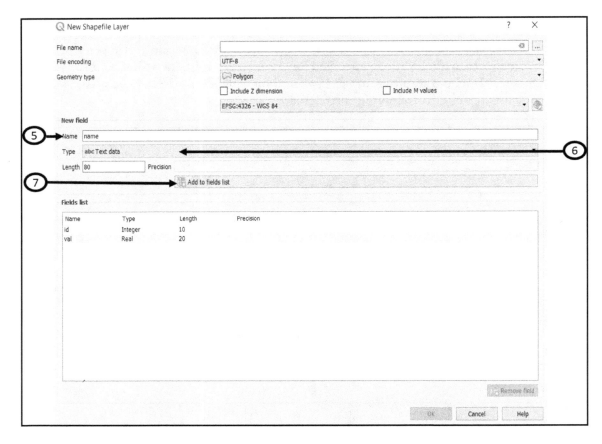

3. We'll see that **name** is added to the **Fields list**. We save this as a polygon shapefile by clicking on the far right of the **File name** and giving it the name Polygon. We need to select the **Geometry type** as Polygon (for line data, select Line), and then save it by clicking **OK**. Now we'll see that this layer is added:

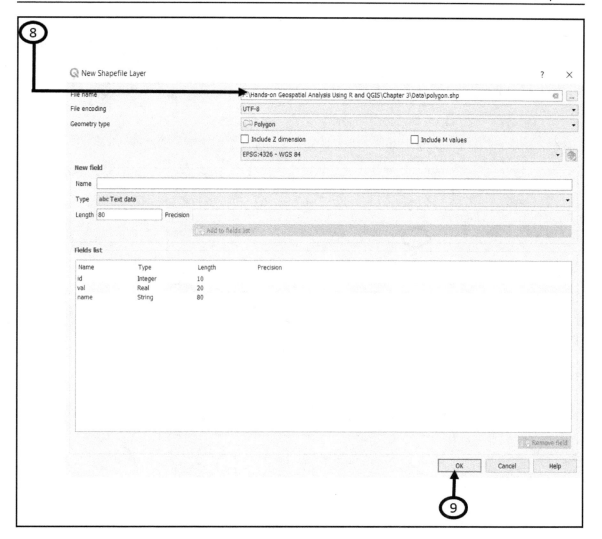

Adding features to vector data

We've created data of the polygon type, and we need to add features now by creating polygons and then filling feature attributes accordingly:

1. The first thing we need to do to add features is enable **Toggle Editing** by clicking on it. Now, we need to draw a polygon; we can do so by clicking on **Add Polygon Features**.

2. Let's create a new polygon now. We need to click on the **Map Display** window and then click on every corner of the polygon we want to create. With every left-click, a node or the corner point of a polygon is created and, with successive clicks, these points are added. When we are done with adding all of the corners, just right-click to finish drawing the polygon. The steps are as follows:

1. Click on any place.
2. Click straight down the first point. We'll see that a straight vertical line has been made.
3. Click a point horizontally away from the second point. Now right-click to stop editing:

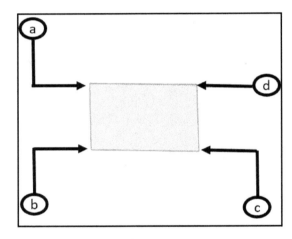

3. Now we'll get a window for filling out the feature attributes information. We fill **id** with 1, **val** with 10, and **name** with first; note that these values are arbitrary and are used for demonstration purposes only:

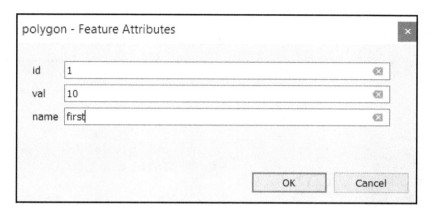

4. We can similarly create multiple polygons and then fill in their attribute information. After we've finished creating features and attributes, we can save this as a shapefile by clicking on **Save As** within **Project**, under the menu bar.

Similarly, we can create line data; in the case of line data, we'll have a single line instead of multiple lines.

Digitizing a map

In this section, we'll learn how to digitize a map, which will allow us to work with this map for further spatial analysis. In doing so, we need to know the coordinates of some of the point locations on this image. We'll use these location coordinates to digitally get coordinates of all points on the image. These points are called **Ground Control Points** (**GCPs**).

Now we'll digitize an image of a district of Bangladesh, Gazipur, using GCPs:

1. Click on **Manage and Install Plugins** under **Plugins**.
2. Now we see the **Plugins** window. We need to write Georeferencer GDAL in the **Search** bar, click on **Georeferencer GDAL**, and then click on **Close**, as shown in the following screenshot:

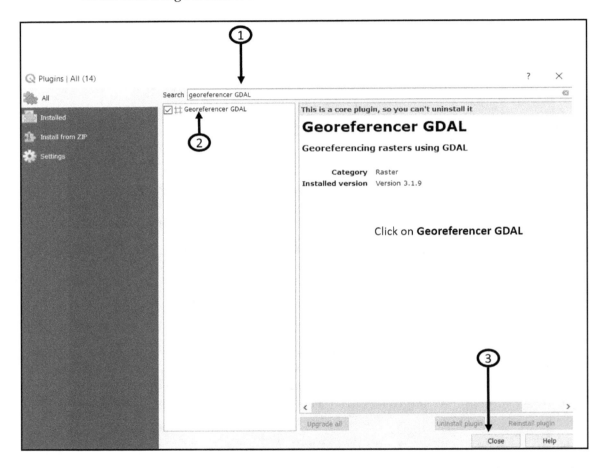

3. We need to open the **Georeferencer** window by clicking on **Georeferencer** under **Raster** from the menu bar, which again shows **Georeferencer**, which we need to click again.
4. A new window for **Georeferencer** will pop up where we can do all of the tasks of digitizing. In this window, we'll now add the image we want to digitize (gazipur.png, found in the Data folder) by clicking on **Open Raster**.

5. We'll get another window for setting up the CRS of this image; we set it to **WGS84**, as you can see from the following screenshot:

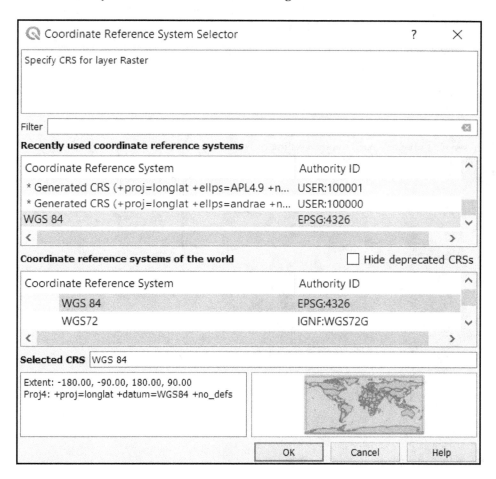

6. We need to add GCPs by clicking on **Add Point** in the **Georeferencer** plugin as follows:

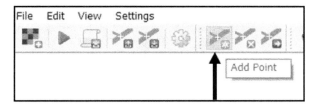

7. If we click on the image inside the **Map Window** inside **Georeferencer**, it will add a GCP there, and then we'll be prompted to enter the coordinates of that point. We'll select four such points in each of the four corners of the image and will use these to digitize the map. After we click on a point, we get the option of entering longitude into the **X / East** box and latitude into the **Y / North** box. After we enter these, we just click **OK** to make our first GCP. We can notice a GCP table, as shown in the following screenshot, in the **Georeferencer** window, which shows different information about this point. Here, creating GCP for one point is shown; using similar steps, we can create three other GCPs:

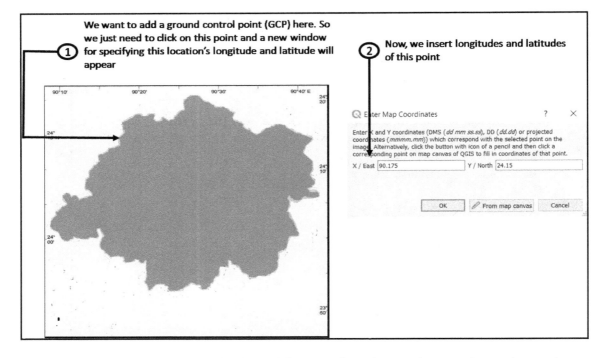

8. After we've selected all of our GCPs, we then change the transformation settings:

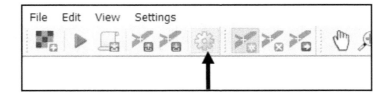

9. We select **Thin Plate Spline** for transformation and **Nearest neighbour** for the resampling method. We then save it as a TIF file, tick the box to load it in QGIS when done, and press **OK**, as can be seen from the following screenshot:

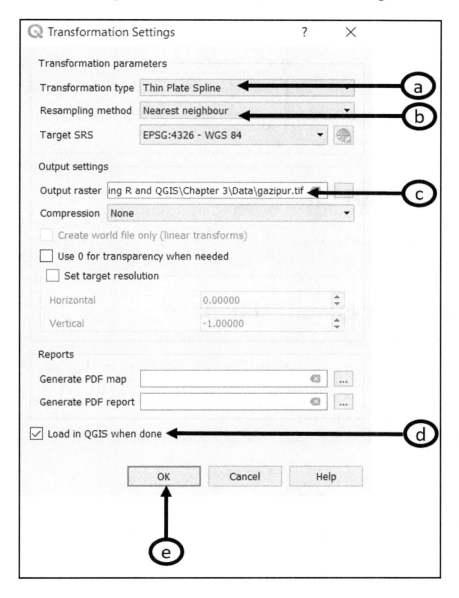

10. We can start georeferencing by clicking on the play button, which will give us a digitized image. Refer to this screenshot:

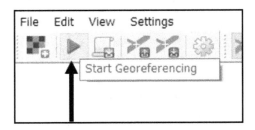

Finally, we have a digitized image.

Working with databases

Using QGIS, we can work with various databases, such as SpatiaLite, PostGIS, MSSQL, and Oracle Spatial. We'll learn here how to create a SpatiaLite database. We'll also learn to import shapefiles into such databases.

Creating a SpatiaLite database

The steps for creating a SpatiaLite database are as follows:

1. We need to right-click on **SpatiaLite** in the **Browser** panel and then click on **Create Database**.
2. We browse to the folder where we want to create the database, give it a name (I named it `Chapter3_test`), and save it.
3. We see an arrow to the left of **SpatiaLite**. Click on this arrow:

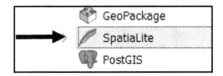

4. We'll see the newly created database under **SpatiaLite**:

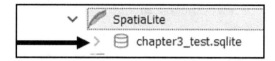

5. Use the **DB Manager** plugin under **Database** to check and manage the database. Now a new window for **DB Manager** will open:

6. Now we can see our database under **SpatiaLite**:

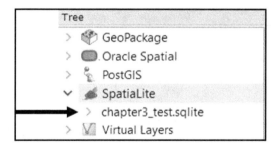

7. We'll see **Table** and **Preview**. By clicking on each of them, we can see relevant information about each of them:

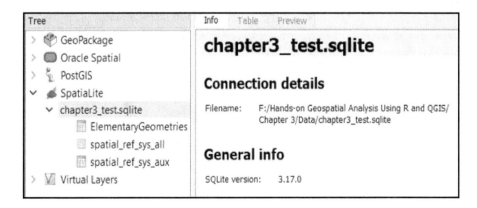

Adding a shapefile to a database

We'll now learn how to add a shapefile to a SpatiaLite database. The steps are as follows:

1. As before, we open **DB Manager**.
2. Click on **Table** and then on **Import Layer/File** to import a shapefile:

3. Import `BGD_adm1.shp`, save it as `bd_division` (under `output` table), and then click **OK** to save it. If everything goes well, we'll see a success notification.
4. This shapefile is added and we can see that this `bd_division` is now being shown under our `chapter3_test.sqlite` database:

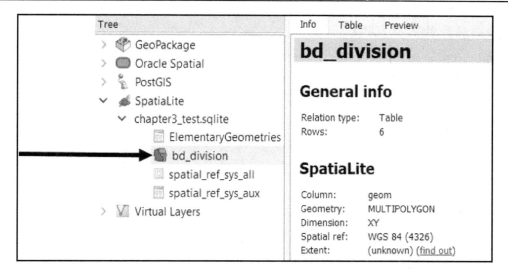

Summary

In this chapter, we've learned how to create vector data and raster data. In doing so, this chapter showed how we can create point, line, and polygon data. Furthermore, it also covered how we can populate different features with attribute values and how we can use the Georeferencer plugin to digitize an image. We ended the chapter by learning how to create spatial databases and how to import shapefiles into them. We've covered just enough to proceed to the next chapters, where we will delve deep into different spatial operations, spatial analysis, and more. We haven't talked in detail about spatial databases and many other operations that could be performed using spatial databases. But the topics covered so far should have equipped you with sufficient resources to dig deeper and, in later chapters, to start applying machine learning models in spatial research cases.

Questions

After completing this chapter, you should now be comfortable in answering the following questions:

- How do you create point, line, and polygon data in QGIS?
- How do you digitize an image?
- How do you create a SpatiaLite database?
- How do you import a shapefile into a SpatiaLite database?

Further reading

We've shown how to create vector data but haven't touched upon the topic of topological error correction. Furthermore, we haven't gone into detail about the various options offered by the GDAL Georeferencer plugin. The book *Mastering QGIS*, by Menke et al, goes into detail explaining these.

Working with Geospatial Data

4

This chapter introduces you to different types of spatial data manipulation in R and QGIS. We'll learn how to merge shapefiles and clip our data to a vector file, differences between shapefiles, how to get the intersection of point data and line data, and how to create a buffer around a feature. We'll use R and QGIS to demonstrate these operations and you'll find out that, for some of these operations, R is more convenient than QGIS, and vice versa. All of these techniques are very helpful to have in a geospatial analyst's toolbox.

After completing this chapter, you'll have hands-on knowledge of the following:

- Combining shapefiles
- Clipping points to the boundary of a shapefile
- Difference
- Intersection between two vector files
- Creating buffers
- Calculating the area of polygons
- Converting vector data types
- Creating statistical summaries of vector files
- Advanced field calculations

Working with vector data in R

Due to the contribution of many developers in the form of R packages, we can now use R as a spatial analysis tool to perform different operations on vector data. To master these, we need to know the basics of spatial data manipulation in R. Some R packages, such as `sp`, `rgdal`, and `rgeos`, will be used frequently to accomplish these tasks.

Combining shapefiles in R

In this example, we'll merge two shapefiles of two districts called Dhaka and Gazipur. Now, `BGD_adm3_data_re` is a shapefile containing all of the districts of Bangladesh as separate polygons. We have the shapefile of Dhaka but not of Gazipur, so we'll create a shapefile of Gazipur before we start merging these two shapefiles. Let's load the required packages now:

```
library(sp)
library(rgdal)
library(maptools)
```

Now, we'll load the `BGD_adm3_data_re.shp` shapefile; as we have seen before, the `NAME_3` field contains the names of the districts. We'll use this field to select the feature that corresponds to the Gazipur district:

```
map_bd = readOGR("F:/Hands-on-Geospatial-Analysis-Using-R-and-
QGIS/Chapter04/Data","BGD_adm3_data_re")
# make a logical vector that has true value only for Gazipur
isGazipur = map_bd$NAME_3 == "Gazipur"
# This will select only those feature(s) whose value equals to "Gazipur"
gazipur = map_bd[isGazipur, ]
# Now, save it using writeOGR()
writeOGR(obj=gazipur, dsn="F:/Hands-on-Geospatial-Analysis-Using-R-and-
QGIS/Chapter04/Data", layer="gazipur", driver="ESRI Shapefile")
```

Now, import this new shapefile and plot it:

```
# Now import gazipur.shp and plot it
gazipur = readOGR("F:/Hands-on-Geospatial-Analysis-Using-R-and-
QGIS/Chapter04/Data","gazipur")
plot(gazipur, col = "blue")
```

This will show the map of Gazipur as follows:

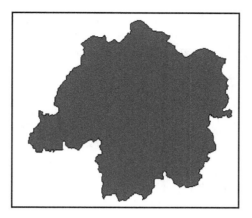

Now, add the map of Dhaka.

We can merge these two `SpatialPolygonsDataFrame` shapefiles using the `union()` function of the `raster` package:

```
# Let's merge these two shapefiles
dhaka_gazipur = raster::union(dhaka, gazipur)
str(dhaka_gazipur, max.level = 2)
plot(dhaka_gazipur)
```

This will result in the following merged map. We can see that these two maps are now merged into a new one:

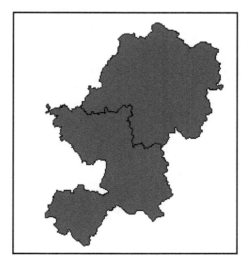

Clipping in R

Here, we'll learn how to clip spatial points to a shapefile. For example, we have many data points over many places of Bangladesh and we want to clip these data points to the shapefile of Dhaka, named `dhaka.shp`. In this example, we have a CSV file, `arbitrary_indicator.csv`, which has two columns, `lon` and `lat`, containing the longitude and latitude of these points, respectively. To achieve our objective, we can take the following steps.

First, read the CSV file and then use `coordinates()` to turn it into a spatial object that R can recognize. In doing so, indicate the longitude and latitude fields followed by ~:

```
points = read.csv("F:/Hands-on Geospatial-Analysis-Using-R-and-
QGIS/Chapter04/Data/arbitrary_indicator.csv")
coordinates(points) = ~ lon + lat
summary(points)
```

Now we can see that `points` is a `SpatialPoints` data type:

```
Object of class SpatialPoints
Coordinates:
        min    max
lon 88.26 92.449
lat 21.43 26.240
Is projected: NA
proj4string : [NA]
Number of points: 261
```

Let's now check how these points look by plotting them on the map of Bangladesh:

```
plot(map_bd, col = "gray", border = "blue", main = "Map of Bangladesh with
arbitrary points plotted")
plot(points, add = TRUE, pch=19, cex=.4, col = "red")
```

The points plotted on the map of Bangladesh appear as follows:

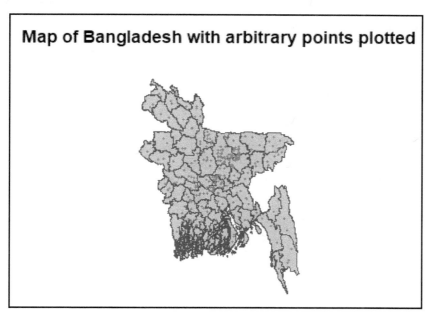

Now, we need do two things: first, make the two layers have the same projection system. Assign the projection of the dhaka layer to points SpatialPoints. Now, subset points by the dhaka layer and that's it, clipping is done in R. We use proj4string() to get the projection system of points and use CRS() to set it to the projection of dhaka, which we get again by using proj4string():

```
proj4string(points) <- CRS(proj4string(dhaka))
points_dhaka <- points[dhaka, ]
```

By adding points_dhaka on the Dhaka map, we can see that points have now been clipped to the Dhaka map:

```
plot(dhaka, col = "gray", border = "blue", main = "Points clipped to the
map of Dhaka")
plot(points_dhaka, add = TRUE, pch=20, cex=1, col = "red")
```

We now see that we have been successful in clipping these points to the shapefile of Dhaka:

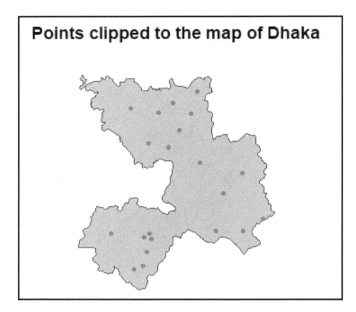

Difference in R

Now, what if we only want the points that are outside Dhaka? That's what the `gDifference()` function under the `rgeos` package in R helps us to do. With `gDifference()`, we provide `points` first, followed by `dhaka`. This is very straightforward to implement:

```
not_dhaka = gDifference(points, dhaka)
plot(map_bd, col = "gray", border = "blue", main = "All the points except
Dhaka's are plotted")
plot(not_dhaka, add = TRUE, pch=20, cex=0.4, col = "red")
```

Here, we have plotted this new `SpatialLines` not_dhaka on the map of Bangladesh to see whether we have been able to keep all of the points except Dhaka's (which are at the middle part of the map of Bangladesh). We'll get a result similar to the following:

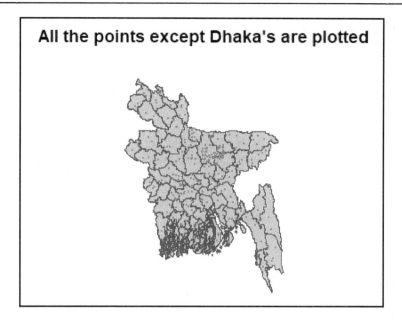

It looks like there are no points on Dhaka, but to be sure, let's also plot these points on the map of Dhaka. If we had been successful in finding the difference, no points should be shown now:

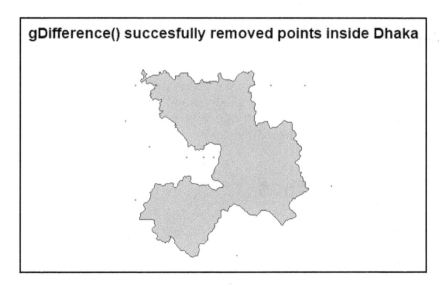

Now we see points around Dhaka and not on Dhaka, as we anticipated, so our work with difference went smoothly.

Area calculation in R

We can very easily calculate the area of all of the features of a polygon using gArea(). We use this to calculate the area, in square kilometers, of different districts in Bangladesh. First, we transform the area using spTransform() to **Universal Transverse Mercator (UTM)** with units in meters; this will make it easy to have the area in square kilometers. We'll use the reprojected map of Bangladesh with UTM in gArea() and use an additional argument, byid = TRUE, so that it computes the area for all unique features (in this case, districts). Now, as gArea() returns in square units of whatever unit in which the map was fetched, we'll get square meters returned by gArea(). So, divide this by 1,000 square to get square kilometers:

```
bd_utm$area = gArea(bd_utm, byid = TRUE) / 1000^2
# We can now check that a new column data with area for each feature has
bee created
head(bd_utm@data)
```

 For more details on using functions such as gUnion(), the gDistance() function of the rgeos package and on vector data manipulation, you can read Chapter 5 of *Learning R for Geospatial Analysis*, by Michael Dorman. For more functionalities provided by QGIS for vector data processing, you can read Chapter 4 of *Mastering QGIS*, by Menke et al.

Working with vector data in QGIS

Now, let's have a look at how we can use QGIS for vector data management. Doing spatial data management in QGIS is not complicated and, once you master it, it's actually enjoyable to perform in QGIS. We'll learn about merging shapefiles, converting polygons into lines and vice versa, clipping, working with difference, buffers, working with intersection, statistical summaries of vector layers, using field calculators, and more.

Combining shapefiles

In many cases, we need to combine multiple shapefiles, for example, when we need to merge the shapefiles of two contiguous areas. We'll merge two contiguous districts in Bangladesh, Dhaka, and Gazipur. We have shapefiles for both and we'll use the **Data Management Tools** of QGIS.

The steps are outlined here:

1. Add the two shapefiles in the QGIS environment. Our display will now look like this. Please note that these two shapefiles are not one but two different files:

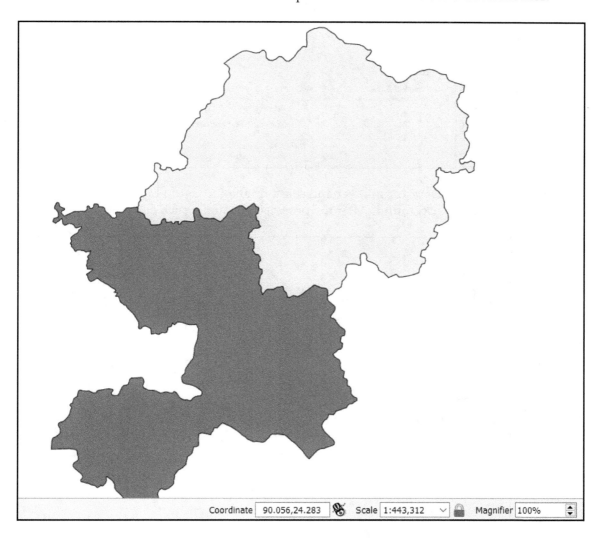

Coordinate | 90.056,24.283 | Scale | 1:443,312 | Magnifier | 100%

2. Click on **Vector** in the toolbar, click or hover on **Geoprocessing Tools**, and then click on **Union...**:

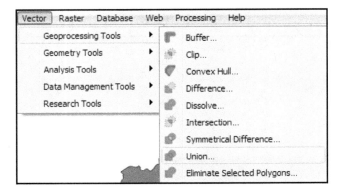

3. Select **gazipur** for **Input layer** and select **dhaka** for **Union layer**, and then click on **Run in Background**. After the processing is done, click on **Close**:

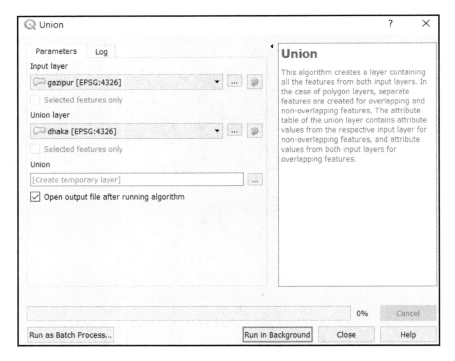

4. We'll get a merged shapefile called Union.

Converting vector data types

We can convert vector data types from one type into another. Here we'll cover how to convert polygon data into line data and vice versa.

Polygons into lines

Now we'll learn how to change polygons into lines using **Geometry Tools**:

1. Open the `polygons.shp` file:

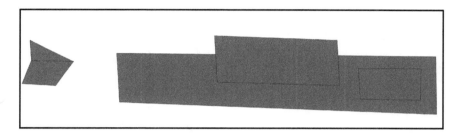

2. Now click on **Vector**, followed by **Geometry Tools**, and then by **Polygons to lines**:

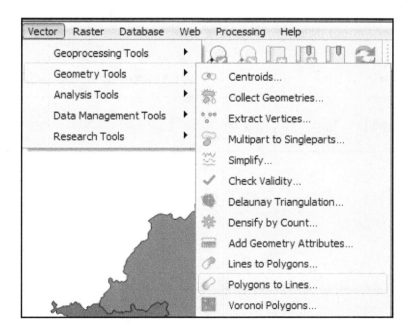

3. We get a window named **Polygons to lines**. Select **polygons** for **Input layer** and then click on **Run in Background**. After the processing is done, click **Close**:

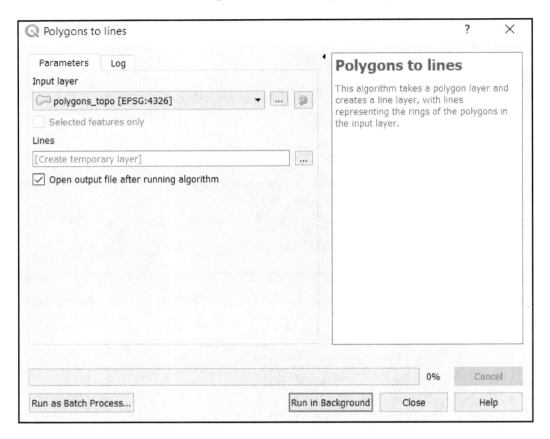

4. We'll get line type vector data converted from the original polygon data:

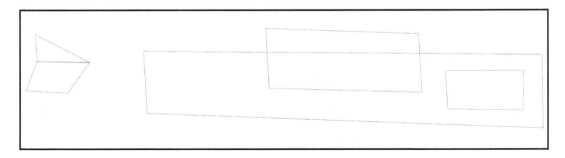

Lines into polygons

Now we'll learn how to do the opposite of what we just did, converting lines into polygons. For this example, we'll be using the `lines.shp` data. The steps are as follows:

1. Load `lines.shp`, which is a vector data of the line type:

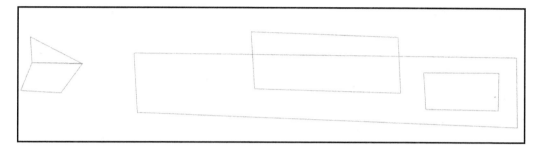

2. Click on **Vector**, followed by **Geometry Tools**, and then by **Lines to polygons**:

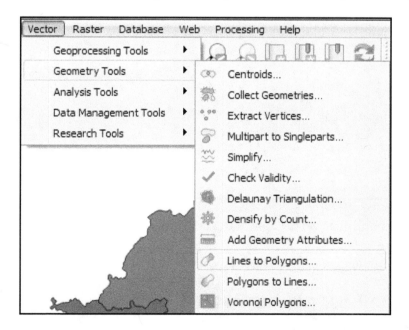

3. Select lines for **Input layer** and then click **Run in Background**:

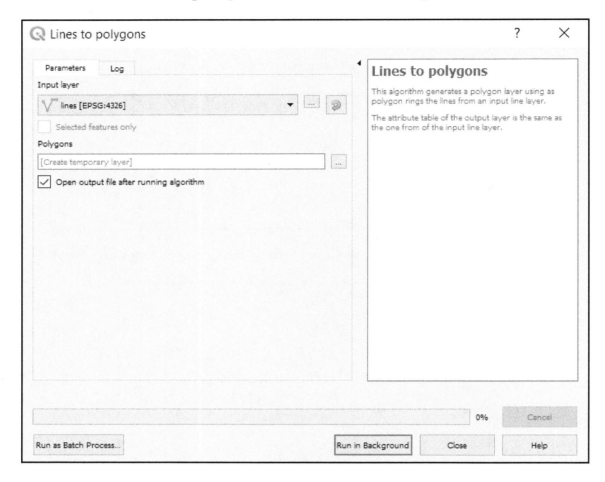

4. We get our lines vector data converted into a new polygon vector:

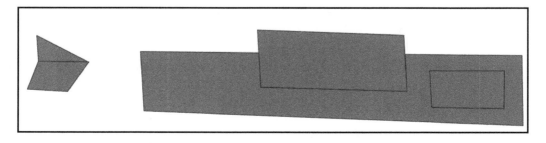

Clipping

Now we have the locations of different areas in Bangladesh where measurements of some variable was taken. We want to plot only those areas that are located inside Dhaka. This can be done very easily using the **Clip** functionality of QGIS. The steps to do so are as follows:

1. Load `BGD_adm3.shp` (loading this is not required; this just adds context), then load `dhaka.shp` followed by the `arbitrary_indicator.csv` file. In the **Layers** panel, ensure that. Our map display should look similar to this:

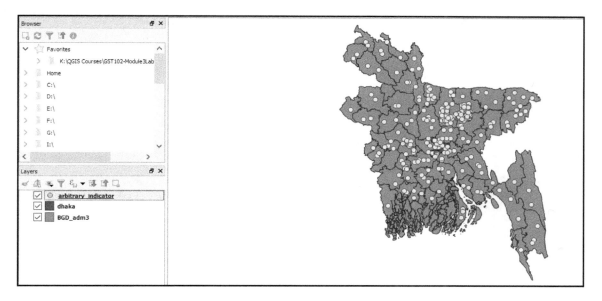

2. Click on **Vector**, followed by **Geoprocessing Tools**, and then click on **Clip**:

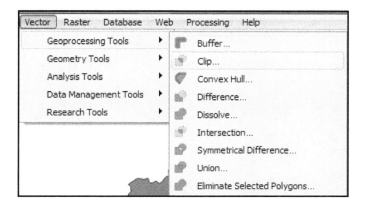

3. Select `arbitrary_indicator` for **Input layer** and select `dhaka` as **Clip layer**, as we want to clip to this area for point data. Click on **Run in Background**. After the processing is done, click **Close**:

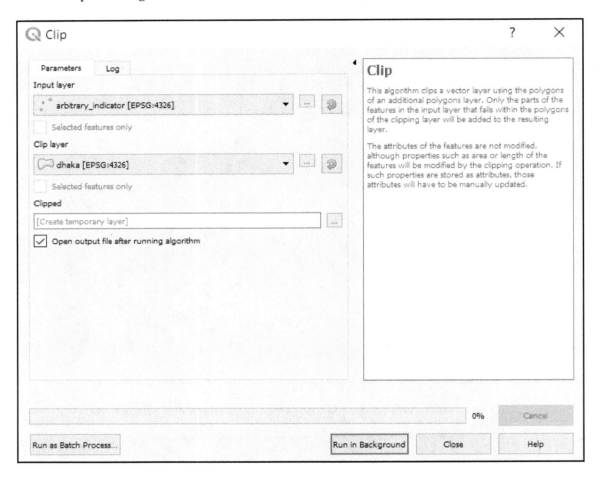

4. We'll see that a new layer, **Clipped**, is created. To see that it really has worked, toggle off the arbitrary_indicator layer (untick it) and we see that points are now clipped to Dhaka:

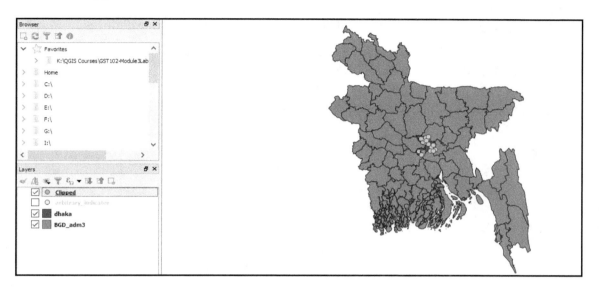

5. To see it more clearly, untick the **BGD_adm3** layer and left-click the **dhaka** layer and select **Zoom to Layer**. Now we can clearly see that points are clipped to Dhaka only:

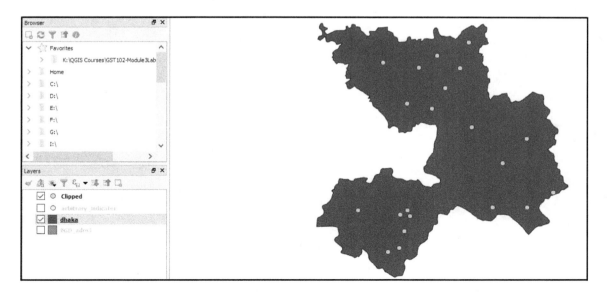

Difference

Now we'll work with the same point data file, `arbitrary_indicator.csv`. We want to plot all of the areas except those that are located inside Dhaka. This can be done very easily using the **Difference** functionality of QGIS. The steps for doing so are as follows:

1. Similar to **Clip**, load `BGD_adm3.shp` (loading this is not required; again, this just adds context), then load `dhaka.shp` ;followed by the `arbitrary_indicator.csv` CSV file.

2. Click on **Vector**, followed by **Geoprocessing Tools**, and then click on **Difference**:

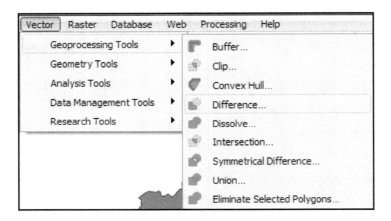

3. Select `arbitrary_indicator` for **Input layer** and select **dhaka** as **Difference layer**, as we want to remove this area for point data. Click on **Run in Background**. After the processing is done, click **Close**:

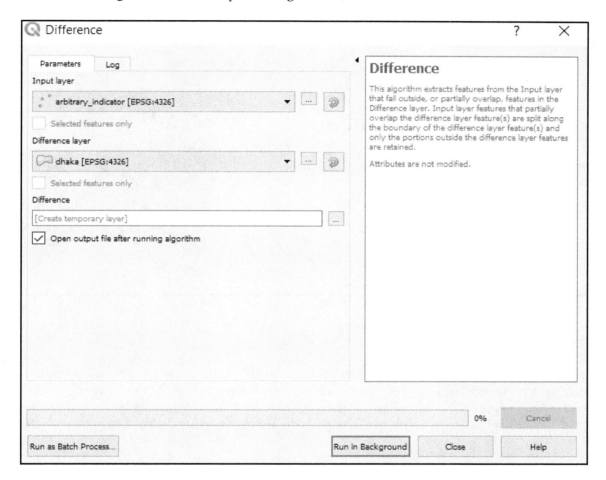

4. We'll see that a new layer, **Difference**, is created. To see that it really has worked, toggle off the **arbitrary_indicator** layer (untick it), and we see that the points selected are all outside of Dhaka:

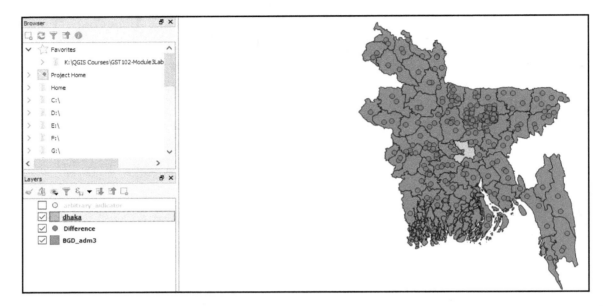

Buffer

Now we'll learn how to create buffer around a vector file. Using this, we can create a buffer around a point, line, or polygon feature. Suppose we have a shapefile of railways in Dhaka named `railway_dhaka.shp`. Now, if we want a buffer around this railway, we can use the buffer feature of the `mmqgis` plugin. The steps are for doing so are described here:

1. Install the `mmqgis` plugin.
2. Load `railway_dhaka.shp`.

3. Select **Create** under **MMQGIS** and then click on **Create Buffers**:

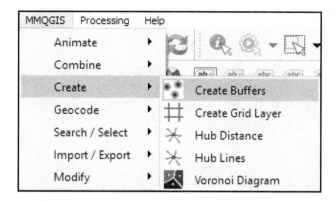

4. Select **railway_dhaka** for **Source Layer**, then put 0.5 for **Fixed Radius** and select **Kilometers** for **Radius Unit**. Select a folder where we want to save this buffer shape and name it `rail_buffer.shp`:

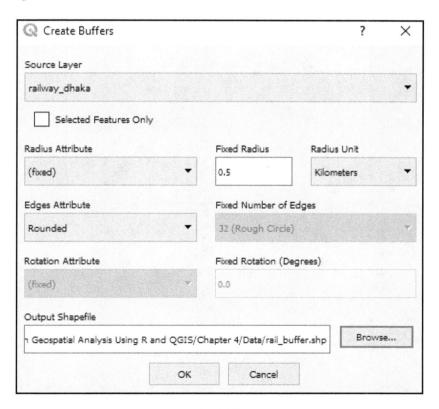

5. Now we'll have a `rail_buffer.shp` file that has a 0.5 kilometer radius around the railway line:

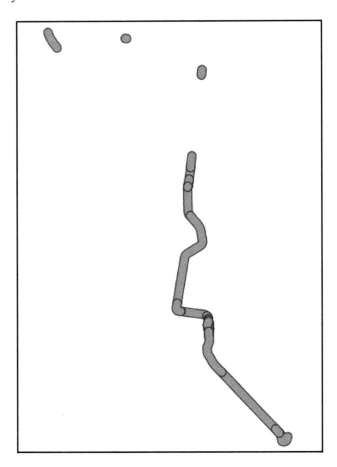

Intersection

Intersection gives us back the intersection of two vector files. So, if we have point vector data and line vector data, the intersection will give all of the point data that intersects with the line data. We will now use food court locations, saved as `food_dhaka.shp`, and see whether there are any food courts within a 0.5 radius of the railway line in Dhaka (contained in `rail_buffer.shp`); we interpret this by seeing all of the points where `food_dhaka.shp` intersects with `rail_buffer.shp`. We can do so by performing the following steps:

1. Load `food_dhaka.shp` and then load `rail_buffer.shp`:

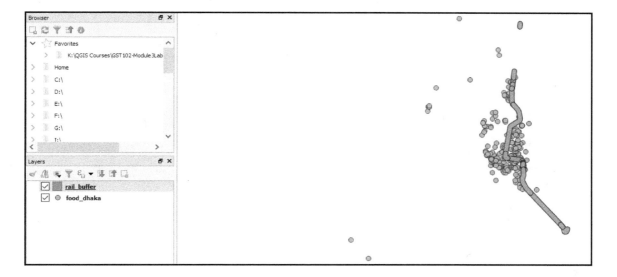

2. Click on **Vector**, then on **Geoprocessing Tools**, and then on **Intersection**:

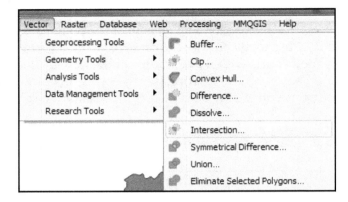

3. Select food_dhaka for **Input layer** and then select rail_buffer for **Intersection layer**. Click on **Run in Background**:

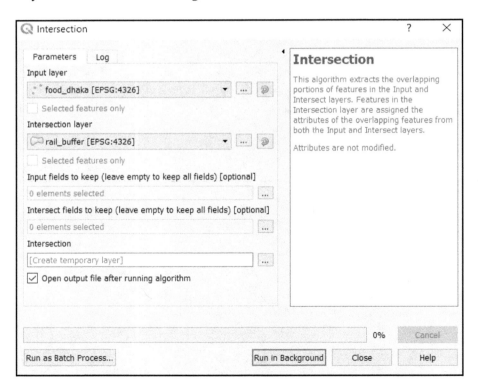

4. We'll get a new layer called Intersection. We can see newly colored points in the map now. Note that all of these points representing food court locations intersect with the 0.5 km buffer around the railway line:

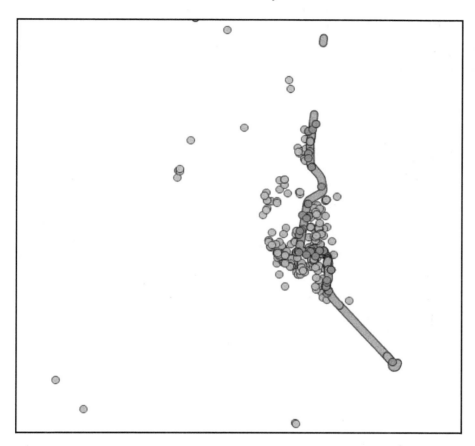

Statistical summary of vector layers

Suppose we have a shapefile that has a numeric field and we want to look at a summary of its statistics, including the count, mean, variance, and quantile. This can be very easily done using the **Statistics** panel of QGIS. We'll now show the step-by-step process of doing this:

1. Open BGD_adm3_data_re.shp.
2. Click on **View** and then click on **Statistical Summary**:

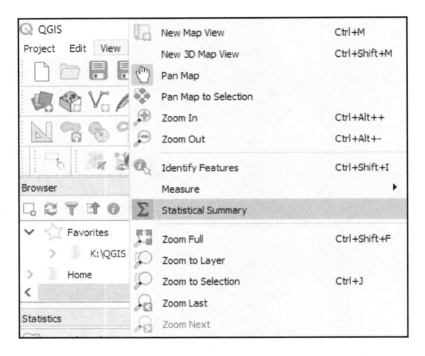

3. We see that the **Statistics** panel is placed between the Browser panel and the Layers panel:

4. Select any field from the drop-down to see the statistics of that field:

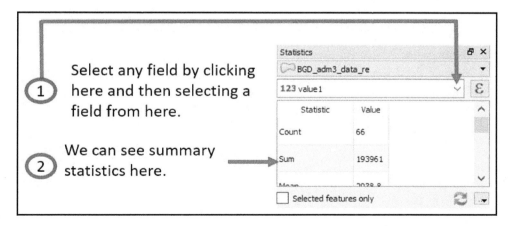

Using field calculators for advanced field calculations

Sometimes, we want to select only those records that meet certain criteria for a particular field or fields or some other constraints. We can very easily select a part of the vector layer that meets our given criteria. In this case, we select a railway polyline from `railway_polyline.shp`:

1. Load the `railway_polyline.shp` shapefile, which is a vector file of railway lines, railway platforms, and other features:

2. Open the attribute table of `railway_polyline.shp` by right-clicking on it in the **Layer** panel.

3. If we scroll down the attribute table, we'll see that there are only two fields and that most of the observations for **railway** don't have any values. This means that only a few of the features actually correspond to the **railway**. It is almost impossible to know all of the values of the **railway** field just by looking at all of these 28,846 features:

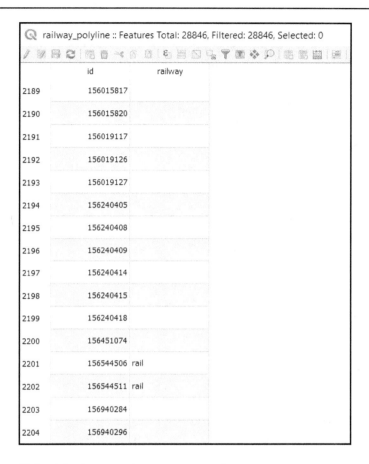

4. Click on **Select features using an expression**:

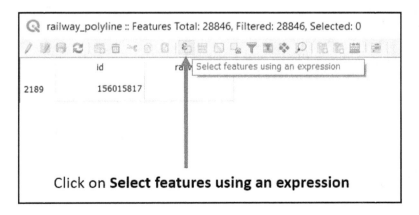

Click on **Select features using an expression**

5. We'll find a new window for Select by Expression. This will allow us to select only those features that have a non-blank value for the railway field and have only those values of the railway field that we want. This window has, in the middle, all of the options for us to use for our expression, including using different field names and values, date and time, different operators, and many other facilities. For example, if we click on Fields and Values, we can see all of the field names. The left part of the window shows the expression; we can also directly edit or write the expression in this part:

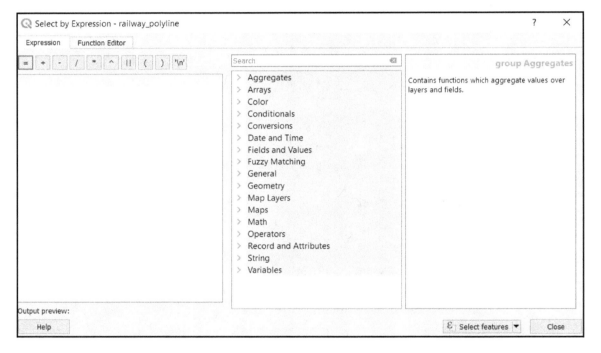

6. We now take the following steps:
    1. Click on **Fields and Values,** which shows all of the fields of this file, then click on **railway**. After that, click on **all unique**; we can then see all of the unique values of **railway**.
    2. Select all of the features that have a value of `railway` that is equal to `rail`. For doing this, we double-click on **railway** under **Fields and Values**. This puts `"railway"` in the left segment of the window in the **Expression**.

3. Click on the = operator and then double-click on **'rail'**. This will give this expression: `"railway" = 'rail'`. Copy this expression, as we'll need it later. Now click on **Select features**. After the processing is done, click **Close**:

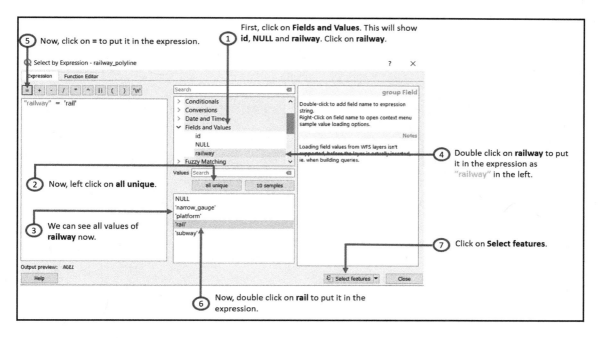

7. We'll see the map again with the features selected where the `railway` field value is equal to `rail`.

8. Left-click on **railway_polyline** and click on **Properties**. Now click on **Source** and then click on **Query Builder**:

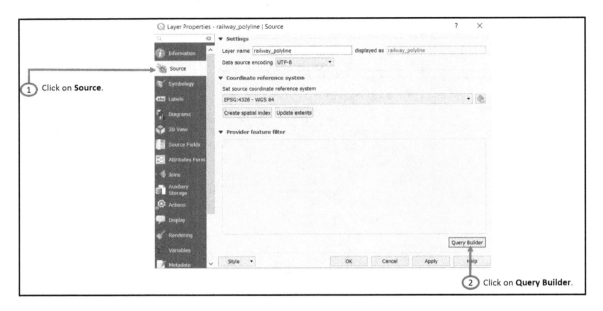

9. Paste `"railway" = 'rail'` in the expression and then press **OK**. This will get us back to the **Layer Properties** window, where we click on **Apply**:

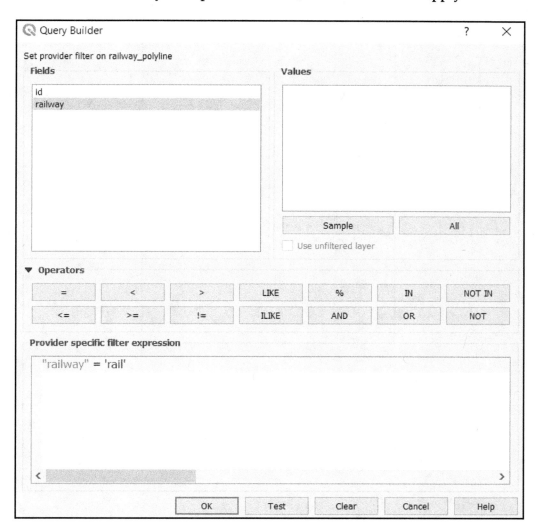

10. We'll see that only a small number of features has been selected. The new map will look as follows:

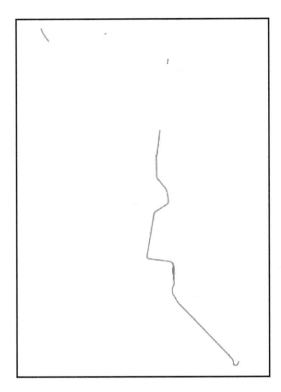

11. Now, if we open up the attribute table again, we'll see that only features with `railway` field values equal to `rail` have been selected. We can now save it as a new layer by left-clicking on **railway_polyline**, clicking on **Save as...**, and saving it as a shapefile.

We have only used the ;+ operator here and have used only one field, but we could have performed a much more complex expression than the one we have just performed. The main purpose was to show you how to use the expression for advanced querying and how, using the different functionalities provided inside **Select by Expression**, we can perform complex queries as required.

Summary

We learned how to use different tools in R and QGIS for working with vector data and modifying or creating new data by doing different operations such as merging, clipping, intersecting, creating buffers, and more. We also learned how to use different spatial queries in QGIS and R. In doing so, we have seen how important it is for two different layers to have the same projection system. We haven't touched upon many other tasks, such as using the rgeos package for calculating distance between geometries using gDistance() or using gUnaryUnion() for taking the union of two shapefiles. With the tools you are equipped with now, you can easily understand other functionalities not covered here and should now be able to perform most of these data manipulation tasks. In doing so, you might already have developed some preferences between QGIS and R for performing each operation. This is actually OK, as, in the next chapter, we'll cover how to blend R and QGIS seamlessly, along with automating our workflow.

5
Remote Sensing Using R and QGIS

In this chapter, we delve deep into working with raster data in both R and QGIS. So far, we have covered vector data, and this chapter will focus on raster data manipulation and analysis. We'll learn how to use raster data provided by different satellites and how to get meaningful information from this data. Going over reading raster data; changing its projection; and visualizing multispectral images to computing slope, aspect, hillshade, and clipping, we'll cover a lot in this chapter, using both R and QGIS. We'll also cover reclassifying rasters, masking raster by a vector layer, and so on.

The topics covered in this chapter are as follows:

- Basics of remote sensing
- Working with raster data in R
- Working with raster data in QGIS

Basics of remote sensing

Remote sensing is the measurement of characteristics or objects of Earth using the reflected radiation or signals that are transmitted from a device and then reflected back to it. For the first case, when the Sun is the primary source of energy, we call this passive remote sensing. By measuring reflected **Electromagnetic Radiation (EMR)**, we try to make sense of the remote sensing data at hand. As different objects have different reflectance across the wavelength of EMR, we can identify different objects by comparing the observed reflectance against the typical reflectance.

Basic terminologies

We'll now briefly look at the following basic terminology of remote sensing:

- **Bands**: Usually, reflected EMR is measured at a different range of wavelengths, which are called bands.
- **DN values**: This reflected EMR is converted into an image where each pixel corresponds to a **discrete number** (**DN**). This number is in the range of 0 to 255 for an 8-bit DN.
- **Radiance**: This is the energy or radiation that reaches the remote sensing instrument.
- **Reflectance**: This is the ratio of reflected energy to total sent energy or radiation.

Remote sensing image characteristics

Remote sensing images can also be classified according to different resolutions:

- **Spatial resolution**: This is the smallest unit area for which we have a DN value. In the case of Landsat images, every DN value corresponds to the reflected energy measured for an area of 30 m x 30 m on the ground.
- **Spectral resolution**: Spectral resolution refers to the ability of a sensor to measure EMR in different wavelength regions. A higher spectral resolution image gives us a higher number of bands and hence more information that can be used for mapping and analyzing different biophysical properties.
- **Radiometric resolution**: This refers to how good the remote sensing is in detecting changes in brightness.
- **Temporal resolution**: This is the time required by a satellite to revisit the same place on Earth. Landsat takes 16 days to revisit the same area on Earth and so its temporal resolution is 16 days.

Atmospheric correction

Electromagnetic (**EM**) energy has to pass through the atmosphere twice in reaching Earth and then reflecting back to sensors positioned in a satellite. In doing so, the EM is scattered and absorbed by aerosols, water droplets, and different gas molecules. So, this atmospheric interaction influences DN, and we need to remove this atmospheric effect to get surface reflectance. Landsat provides an atmospherically corrected image, and we can find this information in the metadata provided with the image. If we get a hazy image or we want to conduct analysis over a large area consisting of multiple images, or if we want to measure change over time, it's better to apply atmospheric correction to the images.

Working with raster data in R

R has some dedicated packages for working with raster data. We can use `raster`, `RStoolbox`, and more to analyze raster data in R. We can use the `readMeta()` function of `RStoolbox` in R to read the metadata of raster data. Let's read the metadata of some Landsat data now:

```
library(RStoolbox)
# Set the working directory
setwd("F:/Hands-on-Geospatial-Analysis-Using-R-and-QGIS/Chapter05/Data")
# Read metadata
meta_data =
readMeta("Mosaic/LC08_L1TP_137043_20180410_20180417_01_T1/LC08_L1TP_137043_
20180410_20180417_01_T1_MTL.txt")
summary(meta_data)
```

This will give us the metadata of this Landsat image, where we can see that it has 11 bands, UTM projection, and `WGS84` as `datum`:

```
Scene:      LC81370432018100LGN00
Satellite:  LANDSAT8
Sensor:     OLI_TIRS
Date:       2018-04-10
Path/Row:   137/43
Projection: +proj=utm +zone=46 +units=m +datum=WGS84 +ellps=WGS84 +towgs84=0,0,0

Data:
                                                      FILES QUANTITY CATEGORY
B1_dn   LC08_L1TP_137043_20180410_20180417_01_T1_B1.TIF      dn    image
B2_dn   LC08_L1TP_137043_20180410_20180417_01_T1_B2.TIF      dn    image
B3_dn   LC08_L1TP_137043_20180410_20180417_01_T1_B3.TIF      dn    image
B4_dn   LC08_L1TP_137043_20180410_20180417_01_T1_B4.TIF      dn    image
B5_dn   LC08_L1TP_137043_20180410_20180417_01_T1_B5.TIF      dn    image
B6_dn   LC08_L1TP_137043_20180410_20180417_01_T1_B6.TIF      dn    image
B7_dn   LC08_L1TP_137043_20180410_20180417_01_T1_B7.TIF      dn    image
B9_dn   LC08_L1TP_137043_20180410_20180417_01_T1_B9.TIF      dn    image
B10_dn LC08_L1TP_137043_20180410_20180417_01_T1_B10.TIF      dn    image
B11_dn LC08_L1TP_137043_20180410_20180417_01_T1_B11.TIF      dn    image
B8_dn   LC08_L1TP_137043_20180410_20180417_01_T1_B8.TIF      dn    pan
QA_dn   LC08_L1TP_137043_20180410_20180417_01_T1_BQA.TIF      dn    qa

Available calibration parameters (gain and offset):
        dn -> radiance (toa)
        dn -> reflectance (toa)
        dn -> brightness temperature (toa)
```

Reading raster data

Using the `raster()` function from the `raster` package, we can read raster data. Now, let's read band 4 of the Landsat image in the `data` folder:

```
library(raster)
band4 =
raster("D:/Spatial_Analysis_1/LC08_L1TP_137044_20170610_20170616_01_T1.tar/
LC08_L1TP_137044_20170610_20170616_01_T1_B1.tif")
band1
plot(band4)
```

The output will look as follows:

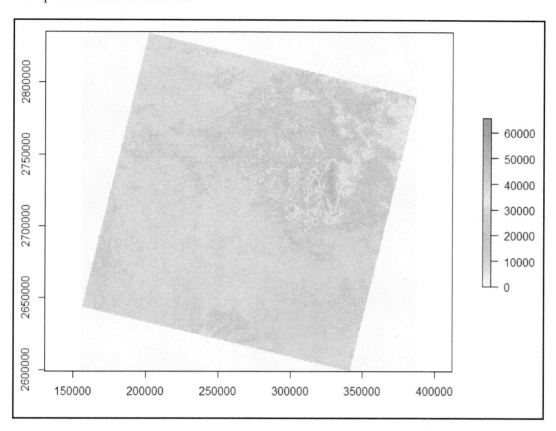

Stacking raster data

We can stack multiple rasters using the `stack()` function. This allows us to read and store multiple bands in R simultaneously. We'll now read and stack bands numbered 1 to 5 using the `stack()` function:

```
band1 =
raster("Mosaic/LC08_L1TP_137043_20180410_20180417_01_T1/LC08_L1TP_137043_20
180410_20180417_01_T1_B1.TIF")
band2 =
raster("Mosaic/LC08_L1TP_137043_20180410_20180417_01_T1/LC08_L1TP_137043_20
180410_20180417_01_T1_B2.TIF")
band3 =
raster("Mosaic/LC08_L1TP_137043_20180410_20180417_01_T1/LC08_L1TP_137043_20
```

```
180410_20180417_01_T1_B3.TIF")
band5 =
raster("Mosaic/LC08_L1TP_137043_20180410_20180417_01_T1/LC08_L1TP_137043_20
180410_20180417_01_T1_B5.TIF")
stacked = stack(band1, band2, band3, band4, band5)
```

Now we can see plots of all of these bands by just plotting the `stacked` variable:

```
plot(stacked)
```

This gives us the following plots of all of the raster data in the `stacked` variable:

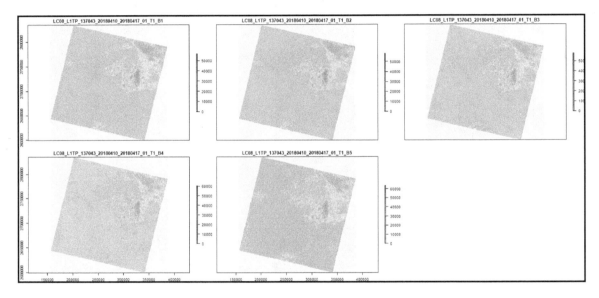

We can also look at the names of all rasters stacked in `stacked` by using the `names()` function:

```
names(stacked)
```

We now can see the names of all bands in `stacked`:

```
[1] "LC08_L1TP_137043_20180410_20180417_01_T1_B1" "LC08_L1TP_137043_20180410_20180417_01_T1_B2"
[3] "LC08_L1TP_137043_20180410_20180417_01_T1_B3" "LC08_L1TP_137043_20180410_20180417_01_T1_B4"
[5] "LC08_L1TP_137043_20180410_20180417_01_T1_B5"
```

We can also use `brick()` for working with multiple rasters at the same time.

Changing the projection system of a raster file

We can change the projection system of a raster file using `projectRaster()` as follows:

```
# Change projection of band4 image from UTM to long lat
band4_ll = projectRaster(band4, crs = '+proj=longlat')
```

We can reproject from longitude and latitude to UTM by using the `projectRaster()` function in the following way:

```
x = " +proj=utm +zone=48 +datum=WGS84 +units=m +no_defs +ellps=WGS84
+towgs84=0,0,0"
band4_utm = projectRaster(band4_ll, crs=x)
```

Now, we'll get the raster back in UTM projection.

False color composite

Images with multispectral layers, such as Landsat images, when opened in any GIS/RS environment, will be rendered as black and white. By assigning different bands of Landsat to either red, green, or blue, we can visualize multispectral layers in different ways. We can assign different bands to red, green, and blue using the `plotRGB()` function:

```
plotRGB(stacked, r = 3, g = 2, b = 1, stretch = "hist")
```

This will now give output that could be useful for vegetation study:

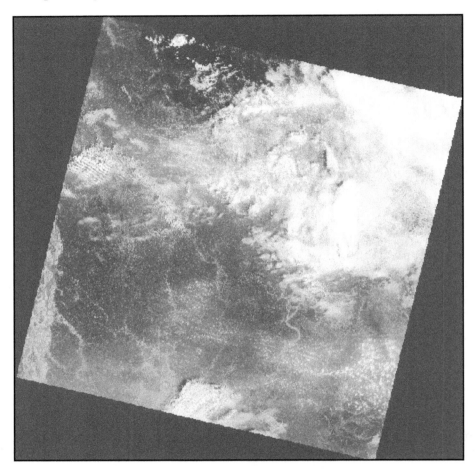

Slope, aspect, and hillshade

We'll now calculate slope and aspect from a **Digital Elevation Model (DEM)**. We'll be using DEM raster data provided by **SRTM**. We now read and plot a DEM using the `raster()` and `plot()` functions:

```
dem = raster("DEM/dem_chittagong.tif")
plot(dem)
```

This DEM looks like this:

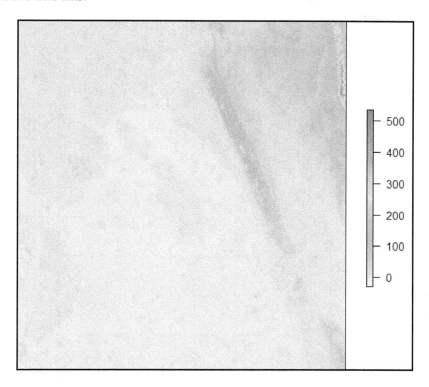

Slope

The slope of a DEM can be calculated in R using `terrain()`, where we need to mention whether we need slope in degrees or percentages by providing values to the `unit` argument:

```
slope = terrain(dem, opt = "slope", unit = "degrees")
plot(slope)
```

This gives us the following output of a slope between 0 and 60:

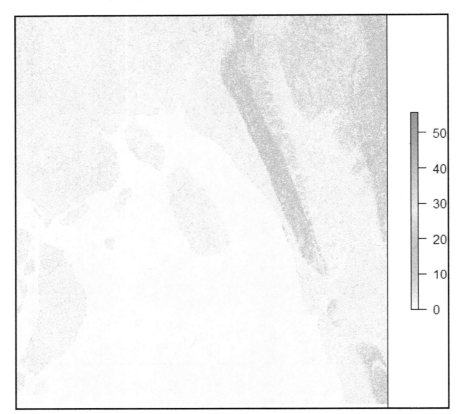

Aspect

The aspect of a DEM can also be calculated in R using `terrain()` and specifying `opt = "aspect"` as follows:

```
aspect = terrain(dem, opt = "aspect")
plot(aspect)
```

This will calculate the aspect or the direction of a slope and gives the following output:

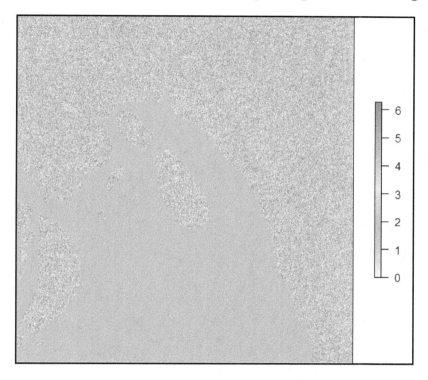

Hillshade

Hillshade illuminates a surface by considering a hypothetical light source and its angle. In R, we need to calculate the slope and aspect before calculating the hillshade. We also have to provide the angle and direction of the sun inside the hillshade() function, as follows:

```
hill_dem = hillShade(slope, aspect, 40, 270)
plot(hill_dem)
```

Now, we'll get the following hillshade output:

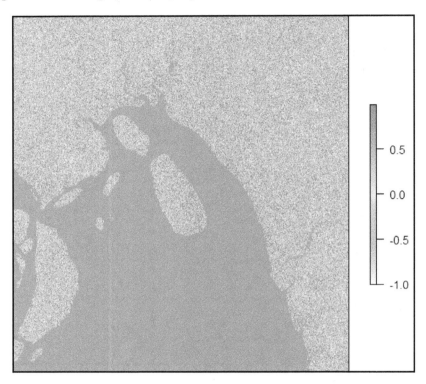

Normalized Difference Vegetation Index (NDVI)

The NDVI is calculated using the near-infrared and visible light reflected by vegetation. The value of the NDVI ranges between -1 and 1. The formula for the NDVI is *(near-infrared - visible)/(near-infrared + visible)*. As for Landsat, band 4 corresponds to near-infrared and band 5 corresponds to visible; this formula, for Landsat, is *(band 4 - band 5)/(band 4 + band 5)*. In R, we can write this as follows and get a plot of the NDVI:

```
NDVI = (band4 - band5) / (band4 + band5)
plot(NDVI)
```

The NDVI will look like the following:

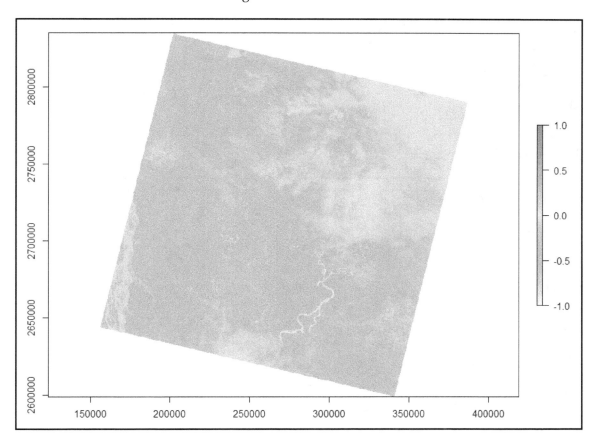

Classifying the NDVI

If we want to reclassify the NDVI into several categories, we can save the classification rule as a matrix and then provide it as the first argument to the reclassify() class. Suppose we want to classify all values between -1 and -0.5 as class 1, values between -0.5 and 0 as class 2, values between 0 and 0.2 as class 3, values between 0.2 and 0.5 as class 4 and values between 0.5 and 1 as class 5.

First, we define a matrix with these rules:

```
rules = c(-1, -0.5, 1, -0.5, 0, 2, 0, 0.2, 3, 0.2, 0.5, 4, 0.5, 1, 5)
class = matrix(rules, ncol = 3, byrow = TRUE)
```

Now, we use the `reclassify()` function to reclassify the NDVI image according to the rules we just defined in `class`:

```
classified_ndvi = reclassify(NDVI, class)
plot(classified_ndvi)
```

The output is as follows:

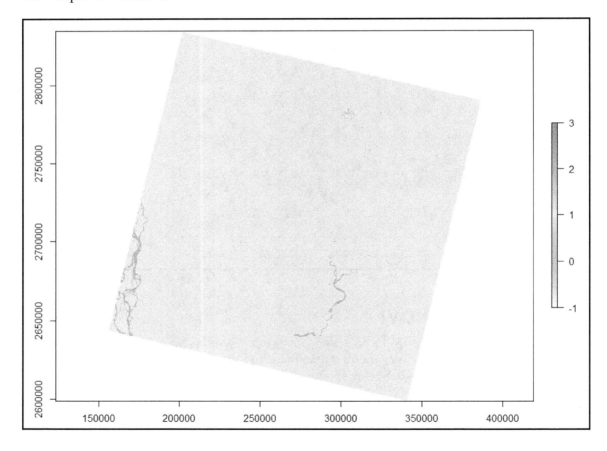

Working with raster data in QGIS

QGIS can be used to perform a number of raster data analysis and management tasks. Having covered how to work with raster data in R, this chapter now looks at how to do some of those tasks, as well as others tasks, in QGIS. Some of the tasks performed in R with raster data can be done faster in QGIS. We'll now have a look at some of these functionalities in QGIS.

False color composite

We'll stack raster images of different bands into a single image with multiple bands using stacking. We'll work with the Landsat images that can be found under the `Mosaic` folder. Inside the `Mosaic` folder, we have a sub-folder, `1`, and inside that, there are multiple images corresponding to different bands. The images are the ones with the `.tif` extension. The steps are as follows:

1. Click on **Raster**, then **Miscellaneous**, and then click on **Merge**, as shown in the following screenshot:

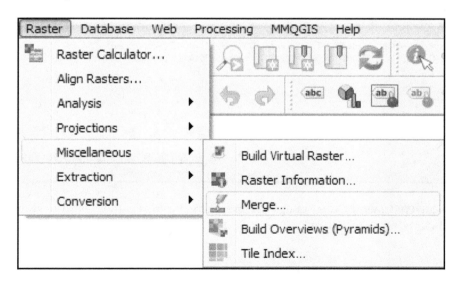

2. We'll find a new window popping out for **Merge**. Tick the checkbox for **Place each input file into a separate band**. Click on **...** to the immediate bottom-right of **Input layers** to select all of the different band images:

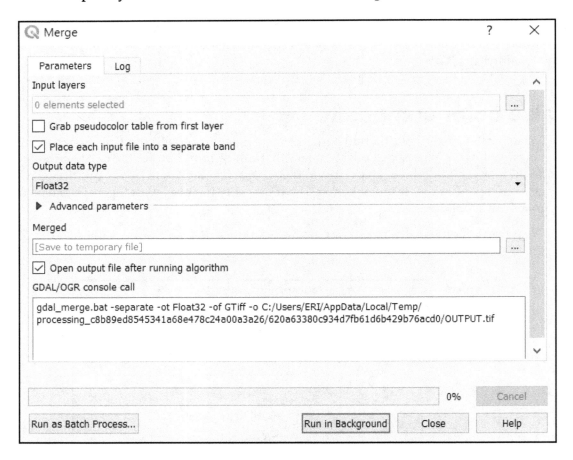

3. Now we have the option to select all of the bands that we want to stack by clicking on **Add files(s)...**:

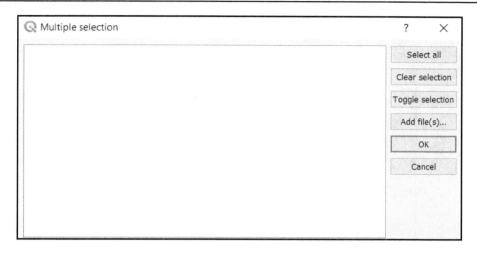

4. Select all of the images with the extension `.tif` or that are TIF images. The path to these files is **All_Data** | **Mosaic** | **1**. Then click **Open**. This will give us a new stacked file named `Merged`:

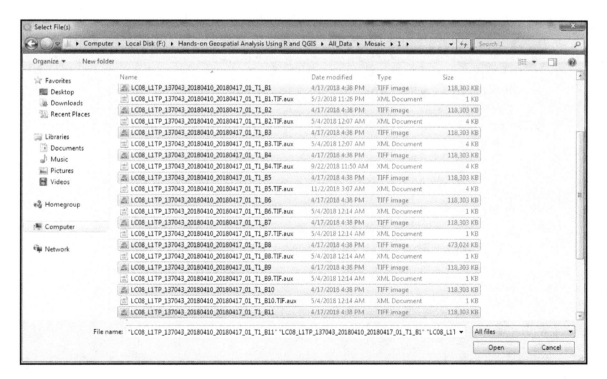

5. We'll see a window where all of the files we selected are shown. Click **OK**:

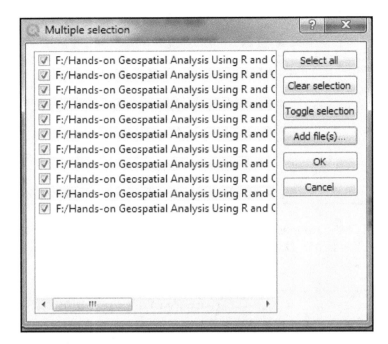

6. Click **Run in Background** and, after it's run successfully, we'll find a new layer called **Merged**.
7. Right-click on the **Merged** layer and select **Properties**, and then select **Style** on the left.

8. Now, to get a colored raster image, we'll need to select a false color composite. We'll select a color composite now by selecting **Band 04** for **Red band**, **Band 03** for **Green band**, and **Band 02** for **Blue band**. Click on **Apply** and then click on **OK**:

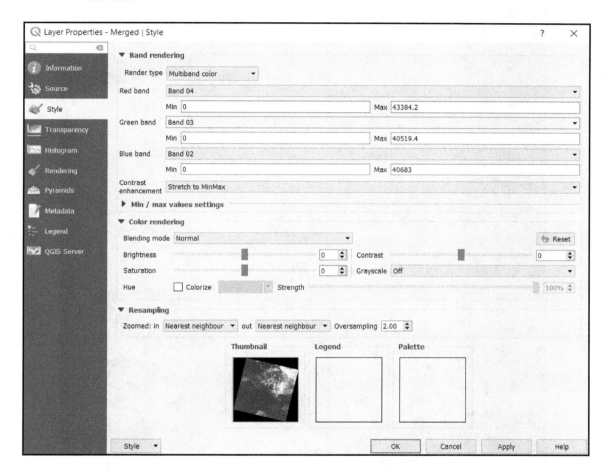

9. We'll get an image that we can save by right-clicking in the **Layers Panel** and then selecting **Save As...**, before saving it as a GeoTIFF file. The image will be similar to the following:

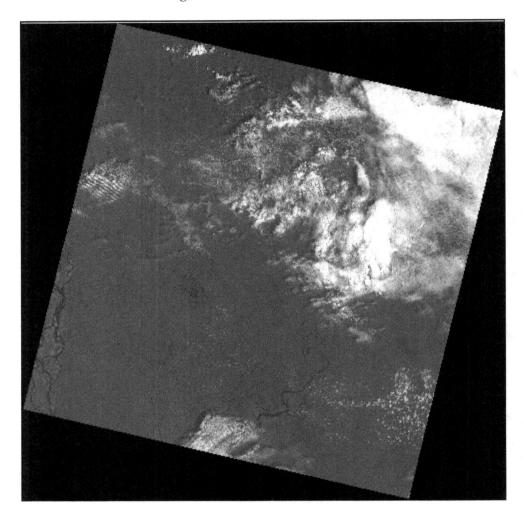

Raster mosaic

Usually, the area of interest for our task will not be covered by a single raster image. As raster data is provided in tiles, by adding two or more tiles, we can get coverage of the complete area of interest. This technique of tiling images together is called mosaic, and in QGIS, we can do this in two ways: using the **Merge** tool and using **Processing Toolbox**.

We have two multi-band raster pictures. We'll learn how to mosaic two raster images using Merge:

1. Add two raster files, 1_chittagong.tif and 2_Khagracchari.tif, from All_Data folder. If there's any difficulty in seeing any of these .tif files, right-click on the .tif file we want to see and click on Zoom to Layer. If we want to see both of these images together, select the Zoom Out button and click on the image. If there is too much zooming in, we can again click on the Zoom In button:

2. Click on **Raster**, then click on **Miscellaneous**, followed by a click on **Merge**:

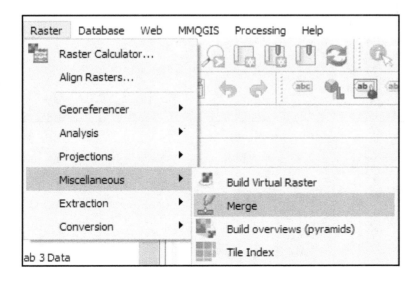

3. We now have a new window for **Merge**. Click on **...** to the immediate bottom-right of **Input layers** to select the files to be merged:

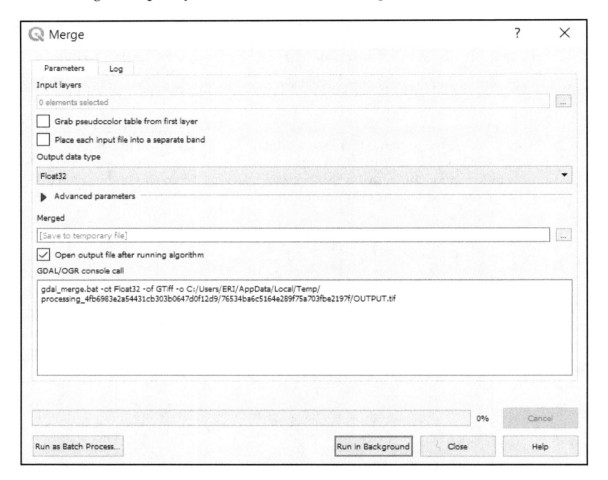

4. Tick the two files, `1_chittagong.tif` and `2_Khagracchari.tif`, and click **OK**:

5. We'll get back to the **Merge** window; click on **Run in Background**:

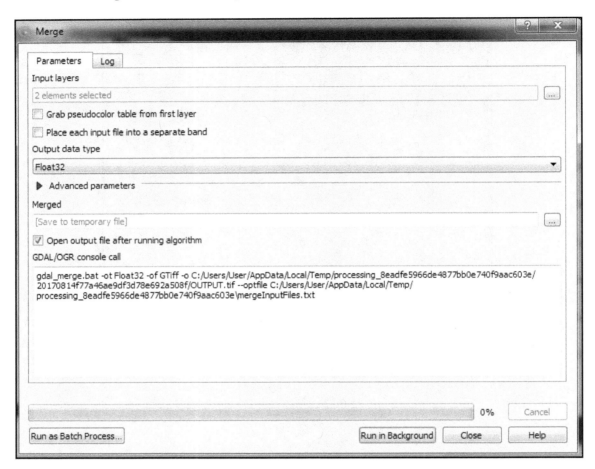

6. Now we see a mosaic of these two files:

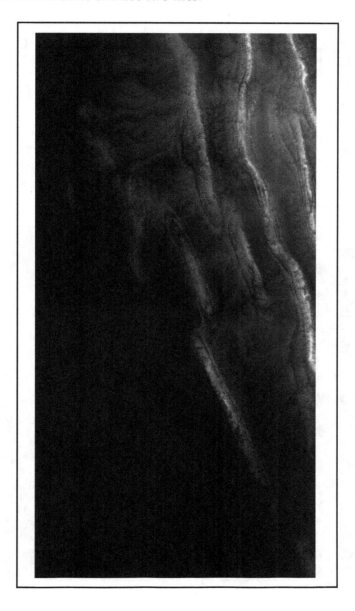

Clip raster by mask layer

Raster images normally come in tiles and cover all of the areas within that tile area. Now, to conduct a research or analysis, we might not need a particular area within this tile. QGIS has a very handy tool for selecting a portion of a raster according to the extent of another vector file.

Now we'll extract a Landsat image according to the shapefile of the Gazipur district:

1. We'll work with band 2 of the Landsat image taken over Bangladesh. Gazipur is located at the bottom of the picture. We now load these two files LC08_L1TP_137043_20180410_20180417_01_T1_B2.TIF in **All_data** | **Mosaic** | **2** and gazipur.shp (in green) in the **All_Data** folder. This would look as follows:

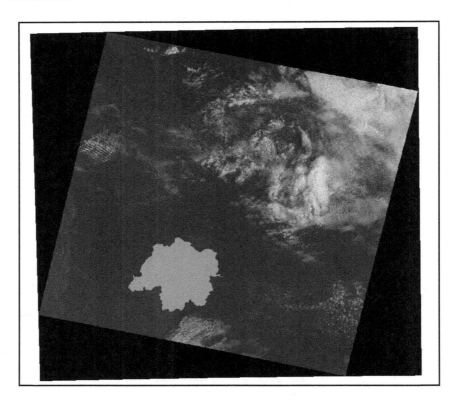

2. We extract this raster file according to the shapefile of Gazipur. To do this, we click on **Raster**, followed by **Extraction**, and then **Clip raster by mask layer**:

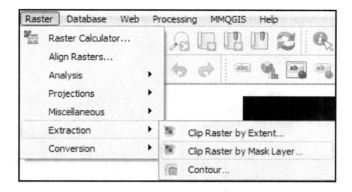

3. We now select the raster image for **Input layer** and **gazipur [EPSG:4326]** as the **Mask layer**. Don't tick any other option and, if any option is already ticked, untick it. Write 255.000000 in **Assign a specified nodata value to output bands [optional]** to remove the black background. Click **Run in Background** now. After the processing is done, click **OK** to return:

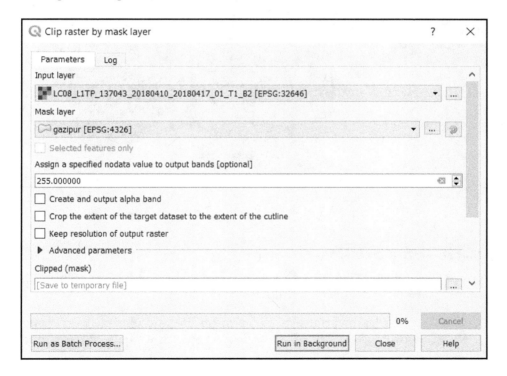

4. We now get a raster clipped to the shapefile of Gazipur only as follows:

Projection system

We can look at the projection system of a raster file by looking at the **Layer Properties** of the raster file. The steps to do so are as follows:

1. Load a raster file, then right-click on a raster file, and then click on **Properties**. We'll be working with 1_chittagong.tif, which we have already used once before.

2. Click on the **Information** tab on the left; we can see **CRS** under **Information from provider**, and we can see spatial extent and spatial resolution:

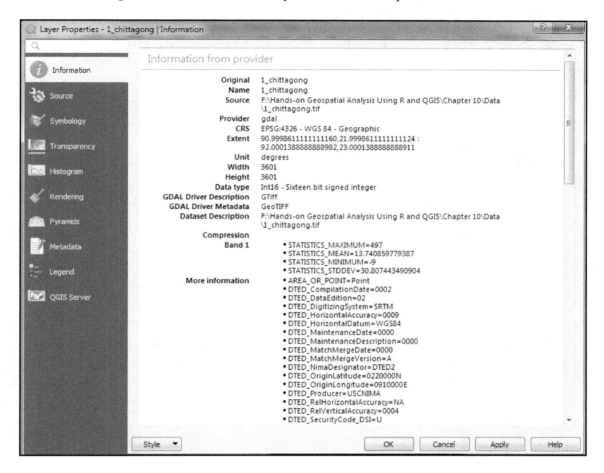

Changing projection systems

We can reproject from one projection system to another in QGIS. This is sometimes necessary when we need to work with two or more rasters that have different projection systems. In such cases, we need to reproject rasters into a common projection system before we can proceed. We'll now reproject a landsat image of Bangladesh from **EPSG:32646** to **EPSG: 3106**. The steps required are described here:

1. Click on **Raster**, then click on **Projections**, followed by a click on **Warp (reproject)**:

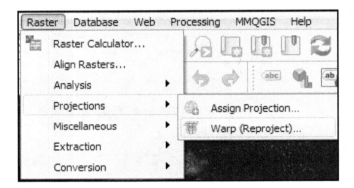

2. Select the band 4 image **LC08_L1TP_137043_20180410_20180417_01_T1_B4** in the folder found at **All_Data** | **Mosaic** | **1** for **Input layer**. Now click on the globe signed box beneath and to the right of **Target CRS**:

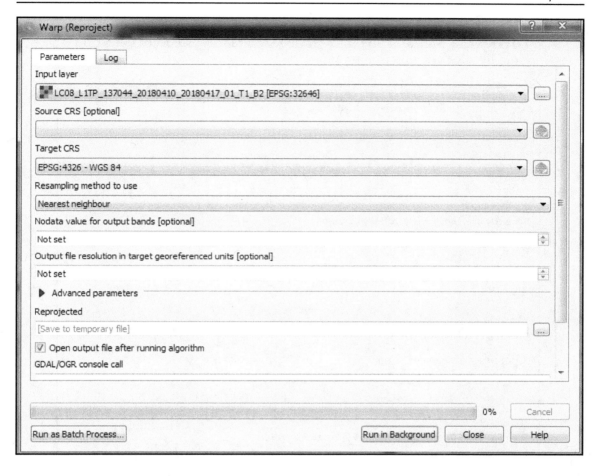

3. Write `EPSG:3106` in **Filter**. After that, click on **Gulshan 303 / Bangladesh Transverse Mercator** under **Coordinate reference systems of the world**, and finally click **OK** to select it:

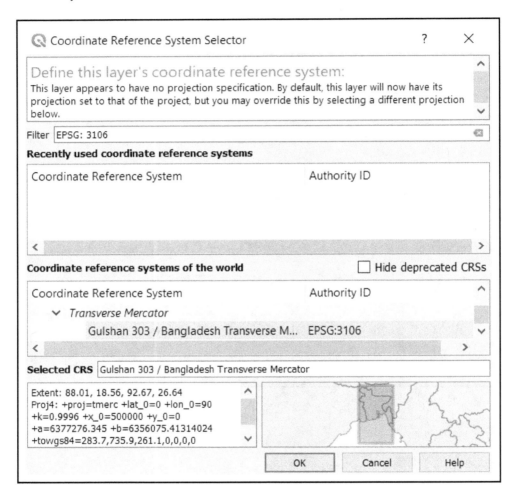

4. We click on **Run in Background** and then click **Close**. We now get a reprojected raster in **EPSG:3106**.

Sampling raster data using points

We can add values from a raster layer to a vector layer. Suppose we have a raster of digital elevation and we have point data in the same extent as covered by the raster. We can get the elevation data for the specific data points sampling raster data. In this case, we'll work with two files: dem_chittagong.tif inside the DEM folder, which contains the digital elevation data in southern parts of Bangladesh; and indicator.shp, which contains some point data (just longitude and latitude values), all of which are within the spatial extent of the raster layer. We want to extract the elevation data from dem_chittagong.tif for points in the indicator.shp shapefile. The steps to accomplish this are as follows:

1. Load the dem_chittagong.tif and indicator.shp files in the All_Data folder. We'll need to install the point sampling tool plugin. First, we need to click on **Plugins** and then click on **Manage and Install Plugins...**. We then click on **Install plugin**. After the installation is completed, click **Close**:

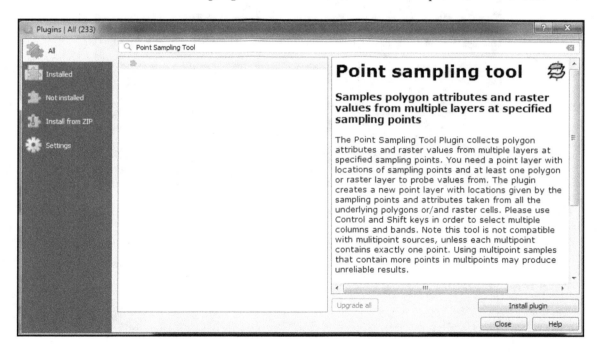

2. These two layers will look as follows:

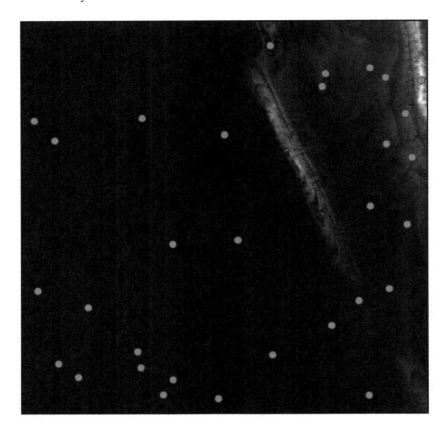

3. We'll now click **Plugins**, then **Analyses**, and then on **Point sampling tool**:

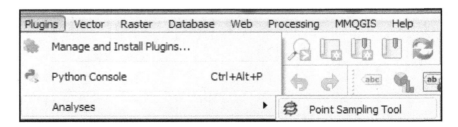

4. Select all of the fields shown under **Layers with fields/bands to get value from:**, then click on the button under **Output point vector layer:** and save it as a shapefile (sampled.shp, in this example). This will create a new shapefile, sampled.shp, which now is the same point vector data as indicator.shp except with a new column with the value of elevation for each of these points. Tick the **Add created layer to the map** field to render it:

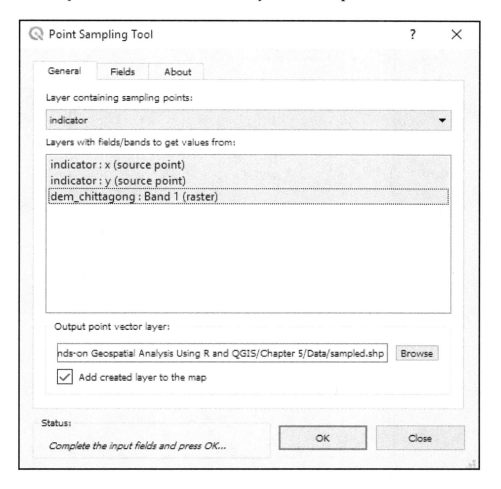

5. Right-click on the newly created sample layer and click on **Open Attribute Table**. We'll now observe that the elevation for the data points are now added as a new field, dem_chitta, along with the two features, x and y, of indicator.shp:

	x	y	dem_chitta
1	91.284999999999997	22.160000000000000	0.00000
2	91.293000000000006	22.119000000000000	0.00000
3	91.347999999999999	22.050000000000001	0.00000
4	91.372000000000000	22.088999999999999	0.00000
5	91.081999999999994	22.699999999999999	8.00000
6	91.031000000000006	22.751000000000001	7.00000
7	91.296999999999997	22.757000000000001	4.00000
8	91.614000000000004	22.940000000000001	19.00000
9	91.742000000000004	22.835999999999999	55.00000
10	91.750000000000000	22.869000000000000	34.00000
11	91.859999999999999	22.882999999999999	45.00000
12	91.897999999999996	22.859000000000002	81.00000
13	91.947000000000003	22.766999999999999	83.00000
14	91.986000000000004	22.850000000000001	403.00000
15	91.900000000000006	22.690000000000001	55.00000
16	91.962999999999994	22.655000000000001	102.00000
17	91.617999999999995	22.152000000000001	0.00000

sampled.shp :: Features Total: 31, Filtered: 31, Selected: 0

Reclassifying rasters

Sometimes, for research purposes, it might be necessary to group numeric values into some categories. For elevation data, we can group the values into some meaningful classes. For example, we can classify elevation data into various categories using processing. Now, we'll reclassify the dem_chittagong.tiff elevation data into four classes. The detailed instructions for doing so are as follows:

1. We load dem_chittagong.tiff first.
2. If we want to recode values from the lowest to 10 as 1, from 10 to 100 as 2, from 100 to 500 as 3, and from 500 upward as 4, we would need to write down the following in a text file:

```
*:10:1
10:100:2
100:500:3
500:*:4
```

3. Save it as value_classification.txt.
4. In the top menu, click **Processing** | **Toolbox**. Write recode in the **Processing Toolbox** that is now positioned to the right of **Map Display**. Now double-click **r.recode - Recodes categorical raster maps**:

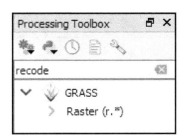

5. Select **dem_chittagong** for **Input layer**. For **File containing recode rules**, browse to the `value_classification.txt` file, which contains classification rules. Click Run. After the processing is done, we'll have a new reclassified raster recoded and we need to click **Close**:

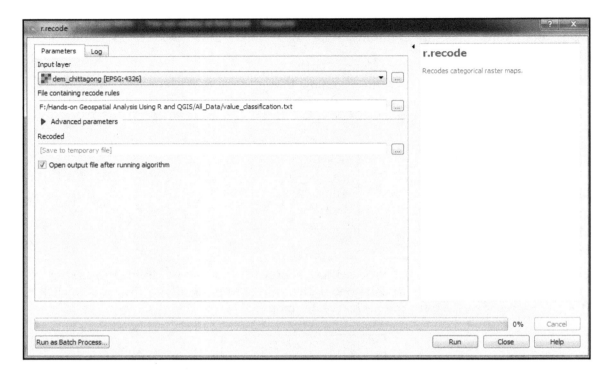

6. We'll now get this reclassified raster of elevation data:

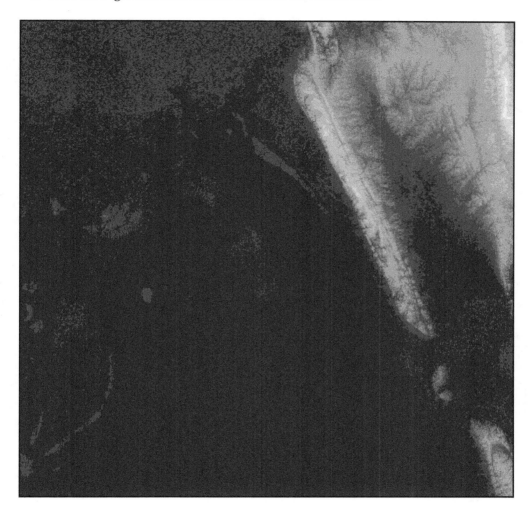

Slope, aspect, and hillshade in QGIS

We can very easily calculate the slope, aspect, and hillshade of a digital terrain model using facilities that can be found under **Raster** | **Analysis**. Slope is calculated as degrees or percentages, aspect gives the slope direction, and hillshade gives an impression of depth. We'll now calculate each of these using QGIS.

Slope

The steps in calculating the slope of a digital terrain model are given here:

1. Click on **Raster** in the top menu. Then, click on **Analysis** and then on **Slope**:

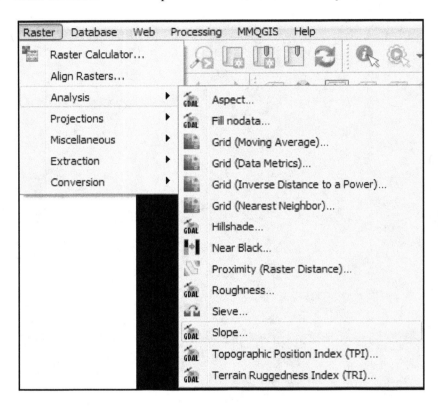

2. Now select `dem_chittagong.tif` for **Input layer** and then click on **Run in Background**:

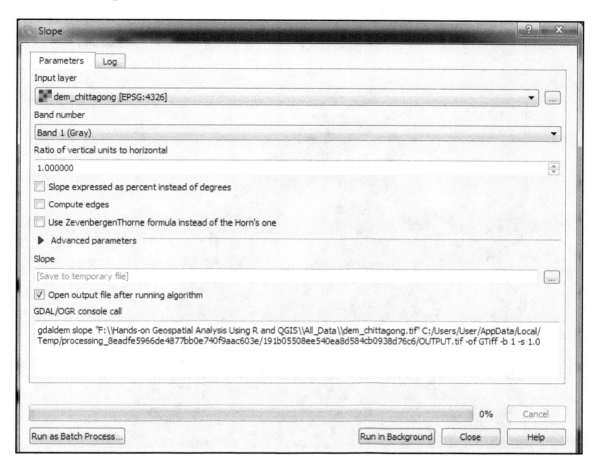

3. We'll now have the following raster:

4. Click on **Raster** in the top menu. Then, click on **Analysis** and then on **Aspect**:

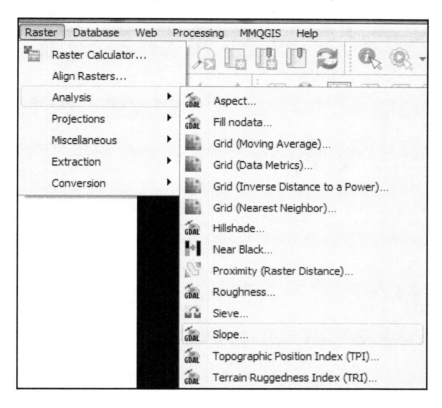

5. Select `dem_chittagong.tif` for **Input layer** and then click on **Run in Background**:

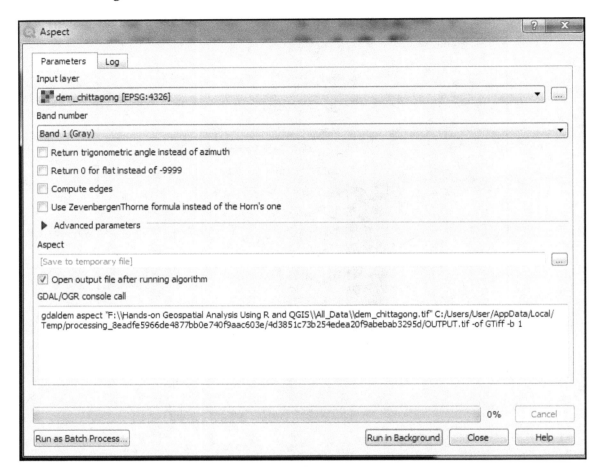

6. We'll now have the following raster:

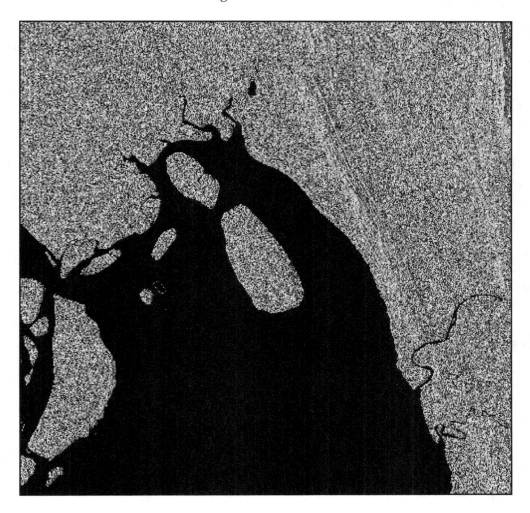

7. By following the same steps of clicking on **Raster** in the top menu, followed by clicking on **Analysis** and then on **Hillshade**, we can get the hillshade.

Summary

In this chapter, we learned how to load, manipulate, and analyze raster data. The things we have learned here will work as the cornerstone for doing complicated raster analysis tasks, such as applying machine learning algorithms, as we'll be discussing in later chapters. We learned how to read and load raster data using the `raster()` function from the `raster` package in R. For multispectral images, such as Landsat images, we have learned how to use `stack()` and how to use `plotRGB()` for false color composite, which gives us colored images. Using the `terrain()` function with different arguments, such as `slope` and `aspect`, we can derive the slope and aspect from DEM data. We can use the `hillShade()` function to generate hillshade in R. Furthermore, if we want to do raster computation, we can just load different bands as variables and then compute these different raster calculations as we have done to calculate the NDVI. In QGIS, we learned how to do almost all of these functionalities. Additionally, we also have learned how to create mosaics of rasters, clip raster data by mask layer, and sample raster data using points.

So far, we have learned how to do basic raster calculations, and so we are now ready to delve deeper. In the next chapter, we'll look at how to do some advanced vector and raster analysis in R and QGIS. We'll look at how to integrate R and QGIS in another chapter.

Questions

Now that you've completed this chapter, you should be ready to answer the following questions:

- How do you load raster data in R and QGIS?
- How do you change projections of rasters in R and QGIS?
- How do you visualize multispectral images in R and QGIS?
- How do you calculate slope, aspect, and hillshade in R and QGIS?
- How do you calculate NDVI in R and QGIS?
- How do you reclassify rasters in R and QGIS
- How do you make mosaics of rasters in QGIS?
- How do you clip rasters by vector data?
- How do you sample raster data using points?

Point Pattern Analysis 6

In this chapter, we'll learn how to use point pattern analysis in R using the `spatstat` package. Point pattern analysis allows us to conduct different statistical analyses on point data. In previous chapters, we learned different ways to manage spatial data—both vector and raster data. This chapter looks at how to conduct statistics-related tasks, such as testing for spatial randomness, spatial logistic regression, and analysis of dependence. In many cases, we might need to draw inferences from our data; using point pattern analysis, we can answer all such queries. This chapter will solely focus on using the `spatstat` package with R only.

In this chapter, we'll cover the following topics:

- Introduction to point pattern analysis
- Analysis of point patterns

Introduction to point pattern analysis

Point pattern data is similar to point data under vector data, which we have already covered. A point pattern dataset gives the location of objects or events of concern in a defined study region. Let's learn some of the terminologies used in point pattern analysis first:

- **Points**: Points are locations in a coordinate system.
- **Events**: If we have data or an observation in a point, we call this an event. These events could represent any spatial objects, such as the location of a crime, a case of disease, a landslide location, or a cell inside the human body. These points are normally in a 2D plane, but they could also be in a 3D plane.
- **Marks**: Marks are another important concept; a mark is simply the attribute information associated with a point; for example, a mark could be a type of crime, a disease type, the intensity of a landslide, or whether a cell is benign or malignant. These marks could be multivariate; so, in the case of a point representing a landslide, the marks could be the land type and settlement status along with the intensity of the landslide.
- **Window**: The finite study area within which the study is defined is called the window.
- **Spatial point pattern**: A set of observed events within the window is called a spatial point pattern.
- **Spatial point process**: This is the stochastic process that generates the events we observe in a window.

Now we are ready to learn about the **planar point pattern** (**ppp**) object in R, which is used for point pattern analysis.

The ppp object

In the spatstat package, for dealing with point pattern process, there is a new data type called the ppp object. This ppp object consists of event coordinates, window, and marks (if there are any).

We need to first install the spatstat package in R and then load it to do point pattern analysis:

```
install.packages("spatstat")
library(spatstat)
```

To create a ppp object, we use the ppp() command. In ppp(x, y, x_dimension_of_window, y_dimension_of_window), we first need to provide the *x* coordinates of the points in the x vector, and the *y* coordinates of the point in the y vector. Then, there's x_dimension_of_window, which is a vector that consists of the coordinates of the left and right corners of the *x* coordinates of the window, and y_dimension_of_window, which consists of the coordinates of the bottom and top corners of the *y* coordinates of the window.

Suppose all of our data points are between 87.71 and 92.79 longitude and between 20.6 and 26.7 latitude. We can now generate 500 different longitude and latitude points ranging in this area using runif(). We'll also need to provide the coordinates for the *x* dimension of the window, which, for convenience, is a little wider than the longitude range: 87.7 and 92.8. Similarly, a range that's a little wider for the *y* dimension of the window is taken. Now let's create a ppp object and plot:

```
bd_ppp = ppp(x = runif(500, 87.71, 92.79), y = runif(500, 20.6, 26.7),
c(87.7, 92.8), c(20.59, 26.71))
bd_ppp
```

We can see that it is a planar object, as shown in the output:

```
Planar point pattern: 500 points
window: rectangle = [87.7, 92.8] x [20.59, 26.71] units
```

Now, we plot this ppp object using plot().

```
plot(bd_ppp)
```

The preceding command gives the following output, with 500 points in the window we specified:

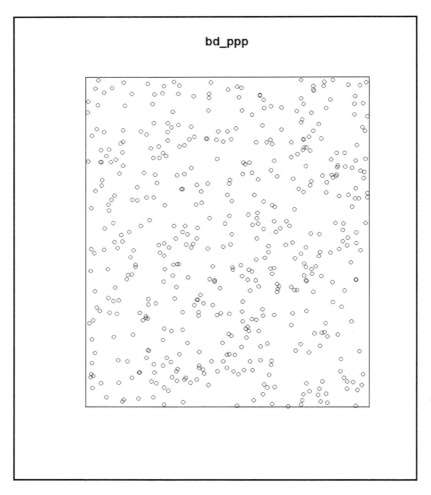

We can also see the basic summary using `summary()` with `ppp`:

```
summary(bd_ppp)
```

This gives summary information about ppp object including the information on number of points, average intensity, spatial extent of the window and the window area:

```
Planar point pattern:   500 points
Average intensity 16.01948 points per square unit

Coordinates are given to 6 decimal places

Window: rectangle = [87.7, 92.8] x [20.59, 26.71] units
Window area = 31.212 square units
```

Creating a ppp object from a CSV file

Suppose we have a CSV file with a location (in longitude and latitude values) and we want to import it into R and make a `ppp` object. Doing so is relatively straightforward—we need to point toward the longitude (*x* coordinates) and latitude (*y* coordinates) values column of the CSV and then use them in `ppt()` as we have done previously. We need to be mindful about the window, so that we don't drop any points.

We have a CSV file, `ppp1`, with the `lon` column containing longitude values and the `lat` column containing latitude values. We'll use `min()` and `max()` to get the minimum and maximum values of longitude and latitude for defining the window containing these values in the following way:

```
# Please update the path to the file ppp1.csv according to where you have
saved # it in your desktop or laptop
location = read.csv("F:/Hands-on-Geospatial-Analysis-Using-R-and-
QGIS/Chapter06/Data/ppp1.csv")
ppp_object = ppp(location$lon, location$lat, c(min(location$lon),
max(location$lon)), c(min(location$lat), max(location$lat)))
plot(ppp_object)
```

Now we get a ppp object:

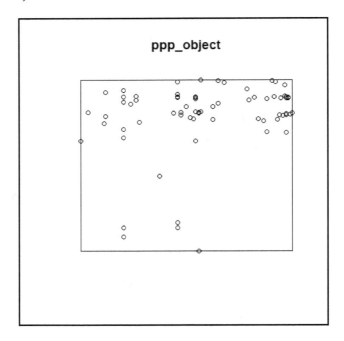

Marked point patterns

In a spatial point pattern, points could carry additional information called marks. These marks could be either continuous or categorical. In the case of a tree, a mark could be the diameter of a tree; in the case of a landslide, a mark could be the time of the landslide. We have to be careful to remember that points are random and so, if the locations are fixed, we don't have valid points.

We'll now import a CSV file where the points are diseases and the marks are the locations of the disease and the disease status (yes or no). In ppp(), to create marks, we have to convert mark data into factors and provide it as a value of the marks argument to the ppp() function. In the following code, we import the data into R as a marked point pattern process and then plot it:

```
# Please update the path to the file ppp2.csv according to where you have
saved # it
marked = read.csv("F:/Hands-on-Geospatial-Analysis-Using-R-and-
QGIS/Chapter06/Data/ppp2.csv")
marked_ppp = ppp(marked$lon, marked$lat, c(min(marked$lon),
```

```
max(marked$lon)), c(min(marked$lat), max(marked$lat)), marks =
factor(marked$value))
plot(marked_ppp, cols = c("green", "red"), pch = c(19, 20), main = "Disease
point pattern")
```

This gives the following marked point pattern process:

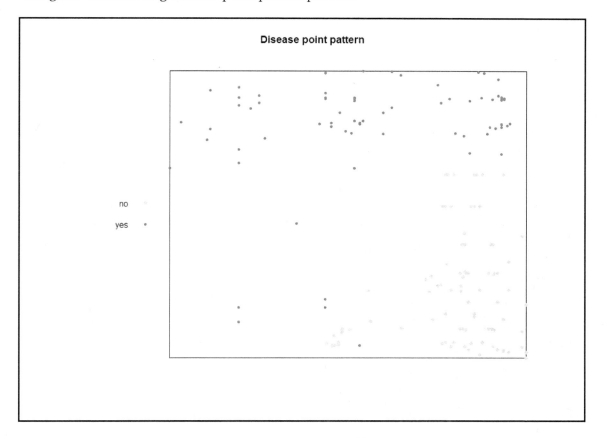

Using `split()`, we can plot points of each type separately. We do this in the following way:

```
plot(split(marked_ppp))
```

This gives us plots for different types:

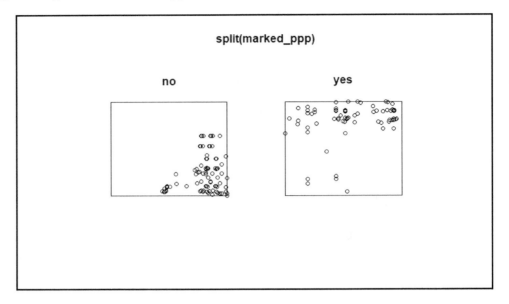

Now let's have a look at this data using `summary()`:

```
summary(marked_ppp)
```

```
Marked planar point pattern:  163 points
Average intensity 128.1813 points per square unit

Coordinates are given to 4 decimal places

Multitype:
     frequency proportion intensity
no          92  0.5644172  72.34775
yes         71  0.4355828  55.83359

Window: rectangle = [89.76, 90.997] x [23.322, 24.35] units
Window area = 1.27164 square units
```

We see a summary of the data in the previous screenshot. Now we can plot separate densities for disease status using `density()` in the following way:

```
plot(density(split(marked_ppp)), main = "Densities for yes and no", ribbon
= FALSE)
```

This gives the following plot of density:

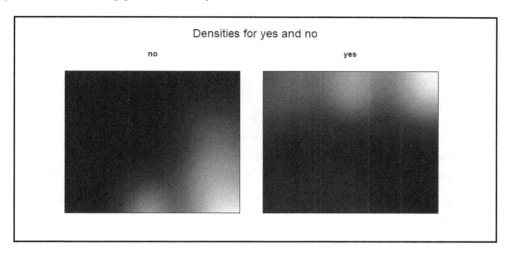

We can also compute relative proportions of intensity using `relrisk()` in the following way:

```
plot(relrisk(marked_ppp), main = "Relative proportions of intensity")
```

This gives the following plot:

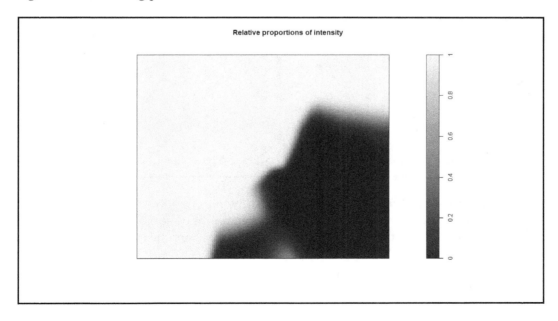

Analysis of point patterns

One of the point pattern processes is called **Complete Spatial Randomness (CSR)**, which means that the probability of finding events at any point is the same everywhere. There are a couple of methods for testing this, which we'll be discussing next.

Quadrat test

In this test, the window is divided into parts and the number of points in each part is computed. If the process is CSR, these numbers in subregions come from a Poisson distribution. Then, using a chi-square test, we can test for CSR.

We'll work with `ppp3.csv`, in this case, and use the `quadrat.test()` function from the `spstat` package to conduct the quadrat test:

```
# update address as necessary
location3 = read.csv("F:/Hands-on-Geospatial-Analysis-Using-R-and-
QGIS/Chapter06/Data/ppp3.csv")
ppp_object3 = ppp(location3$lon, location3$lat, c(min(location3$lon),
max(location3$lon)), c(min(location3$lat), max(location3$lat)))
quadrat.test(ppp_object3)
```

This gives the following output:

```
        Chi-squared test of CSR using quadrat counts
        Pearson X2 statistic

data:   ppp_object3
X2 = 176.54, df = 24, p-value < 2.2e-16
alternative hypothesis: two.sided

Quadrats: 5 by 5 grid of tiles
Warning message:
Some expected counts are small; chi^2 approximation may be inaccurate
```

As `p-value` is much lower than 0.01, we can reject the null hypothesis that the points are completely spatially random, as we can see that there might be some spatial pattern in these events. Also, note that there are some cells or parts where the counts are really small, indicating potential inaccuracies in the chi-square estimate.

We can check whether there is a clustered pattern by using `alternative = "clustered"` inside `quadrat.test()`, as follows:

```
quadrat.test(ppp_object3, alternative = "clustered")
```

This gives the following result:

```
            Chi-squared test of CSR using quadrat counts
            Pearson X2 statistic

data:  ppp_object3
X2 = 176.54, df = 24, p-value < 2.2e-16
alternative hypothesis: clustered

Quadrats: 5 by 5 grid of tiles
Warning message:
Some expected counts are small; chi^2 approximation may be inaccurate
```

As `p-value` is way lower than 0.01, we can reject the null hypothesis of CSR in favor of clustered point pattern process.

G-function

`G-function` is a useful function for describing clustering in point patterns. In this process, the distance of the nearest neighbor for each event is calculated and then the cumulative distribution of the nearest neighbor distance is used to give the probability of an event occurring within a distance, *d*. We can use `G-function` in R using `Gest()`. We can get the range of estimates for *d* values using `envelope()`, which presents them as a shaded region. We'll also need to adjust for the events near the window as these points might not give an unbiased nearest-neighbor distance—what can also be given as an argument to `envelope()`. If the data is over the envelope or the shaded region, that would indicate clustering:

```
ppp_gf = envelope(ppp_object3, Gest, correction = "border")
plot(ppp_gf, main = "G-function")
```

This gives us the following plot:

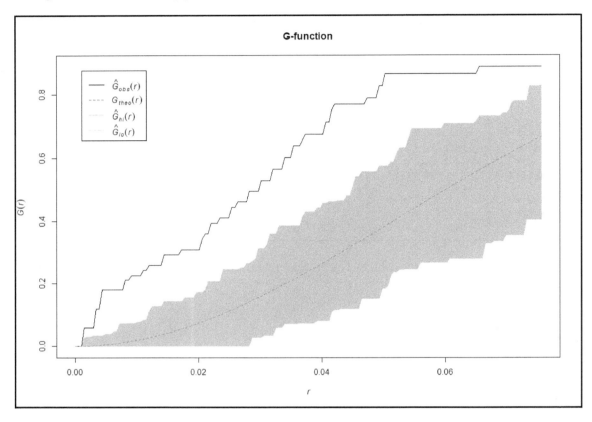

We can see that our observed cumulative distribution is above the high and low ranges of the theoretical cumulative distribution, therefore suggesting clustering.

K-function

`K-function` is another useful function for clustering point patterns. For different values of distance, *r*, the function takes the average number of events in a radius, *r*, around every event: *K(r)*. Similar to `G-function`, we can also use `envelope()` to estimate the upper and lower values for every *r* under consideration. Similar to before, if, for the observed data, *K(r)* is greater than the upper limit of the range, there is the possibility of clustering:

```
ppp_gf = envelope(ppp_object3, Kest, correction = "border")
plot(ppp_gf, main = "K-function")
```

This gives us the following plot:

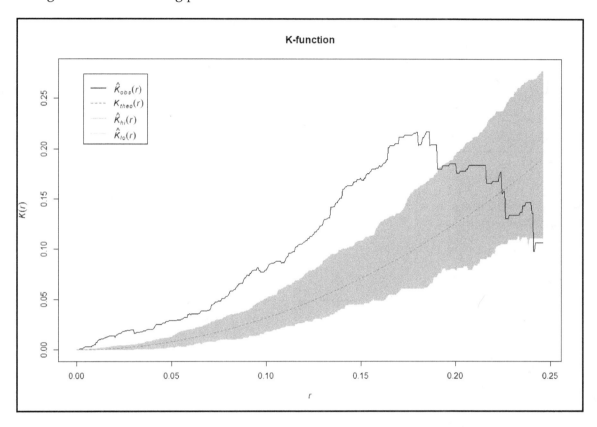

For many points, the observed $K(r)$ is higher than the envelope or above the theoretical and the upper range of this value, so we can say that there is clustering.

If we have a mark and want to test the null hypothesis of CSR, we can use maximum absolute deviation with `mad.test()`, as follows:

```
mad.test(ppp_object3, Kest)
```

This gives us the following output:

```
Generating 99 simulations of CSR ...
1, 2, 3, 4, 5, 6, 7, 8, 9, 10, 11, 12, 13, 14, 15, 16, 17, 18, 19, 20, 21, 22, 23, 24, 25, 26, 27, 28, 29, 30, 31, 32, 33, 34, 35, 36, 37, 38, 39, 40, 4
1, 42, 43, 44, 45, 46, 47, 48, 49, 50, 51, 52, 53, 54, 55, 56, 57, 58, 59, 60, 61, 62, 63, 64, 65, 66, 67, 68, 69, 70, 71, 72, 73, 74,
75, 76, 77, 78, 79, 80, 81, 82, 83, 84, 85, 86, 87, 88, 89, 90, 91, 92, 93, 94, 95, 96, 97, 98,  99.

Done.

        Diggle-Cressie-Loosmore-Ford test of CSR
        Monte Carlo test based on 99 simulations
        Summary function: K(r)
        Reference function: theoretical
        Alternative: two.sided
        Interval of distance values: [0, 0.246]
        Test statistic: Integral of squared absolute deviation
        Deviation = observed minus theoretical

data: ppp_object3
u = 0.0064943, rank = 1, p-value = 0.01
```

As p-value is lower than 0.01, we can reject the possibility of CSR.

We can also use the sum of the squared distance between different simulated K-functions to test the null hypothesis of CSR. We can do this using dclf.test():

```
dclf.test(ppp_object3, Kest)
```

This gives the following output:

```
Generating 99 simulations of CSR ...
1, 2, 3, 4, 5, 6, 7, 8, 9, 10, 11, 12, 13, 14, 15, 16, 17, 18, 19, 20, 21, 22, 23, 24, 25, 26, 27, 28, 29, 30, 31, 32, 33, 34, 35, 36, 37, 38, 39, 40, 4
1, 42, 43, 44, 45, 46, 47, 48, 49, 50, 51, 52, 53, 54, 55, 56, 57, 58, 59, 60, 61, 62, 63, 64, 65, 66, 67, 68, 69, 70, 71, 72, 73, 74,
75, 76, 77, 78, 79, 80, 81, 82, 83, 84, 85, 86, 87, 88, 89, 90, 91, 92, 93, 94, 95, 96, 97, 98,  99.

Done.

        Diggle-Cressie-Loosmore-Ford test of CSR
        Monte Carlo test based on 99 simulations
        Summary function: K(r)
        Reference function: theoretical
        Alternative: two.sided
        Interval of distance values: [0, 0.246]
        Test statistic: Integral of squared absolute deviation
        Deviation = observed minus theoretical

data: ppp_object3
u = 0.0064943, rank = 1, p-value = 0.01
```

As p-value is 0.01, we can reject the null hypothesis of CSR at 1% level for this test also.

L-function

L-function is a simple transformation of K-function and can be used as an alternative to K-function. Let's code it:

```
ppp_lf = envelope(ppp_object3, Lest, correction = "border")
plot(ppp_lf, main = "L function")
```

We now see a similar plot to K-function:

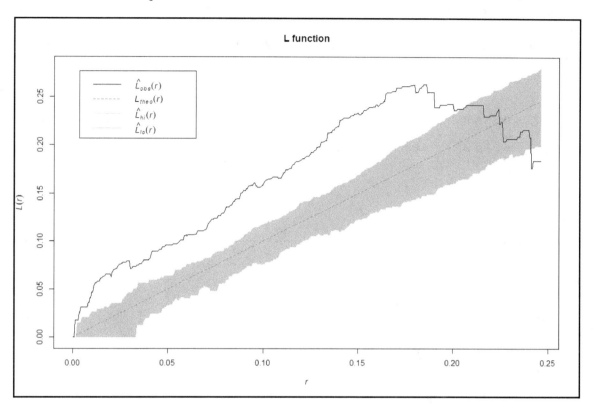

We can use mad.test() and dclf.test() with L-function for the null hypothesis of CSR as we did with K-function.

Spatial segregation for a bivariate marked point pattern

For bivariate marked point pattern processes such as case and control events, a null hypothesis could be the probability of disease being the same in every place against the alternative of it not being the same in all places. The spatialkernel package allows us to conduct significance tests for spatial segregation. First, we'll need to choose a bandwidth for kernel smoothing. Using the spseg() function, we can get it in the following way:

```
bandwidth = spseg(marked_ppp, h = seq(0, 100000, by = 50), opt = 1)
```

Now, we are looking for a bandwidth between 0 and 1,00,000 meters with a 50-meter step. Here `opt = 1` tells `spseg()` to look for the best bandwidth. This returns a list and the best bandwidth is stored in the `hcv` element, which we can access using `bandwidth$hcv`. Now, to do the test, we'll again use `spseg()`, but with some different and some new values for the arguments—now h will be `bandwidth$hcv`, `opt` will be 3 for conducting the test, and the `ntest` argument with a value of `100` will be added so we can do 100 simulations:

```
simulations = spseg(pts = marked_ppp, h = bandwidth$hcv, opt = 3, ntest =
100, proc = FALSE)
```

Now we can plot the probability of disease, `"yes"`, by using `plotmc()` in the following way:

```
plotmc(simulations, "yes")
```

We get the following output:

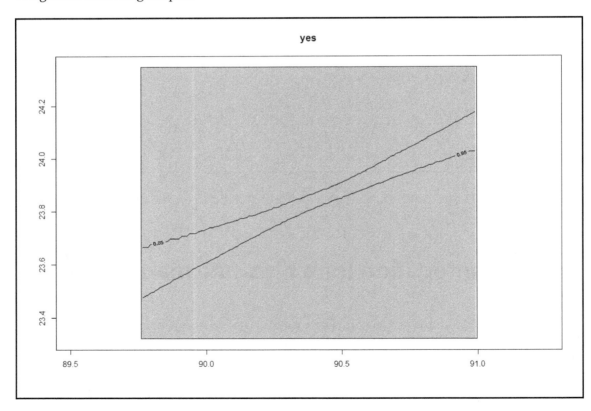

Summary

In this chapter, we have learned about the basics of point pattern processes, including what points, events, marks, and windows are, as well as the `ppp` object. Using the `spatstat` package of R, we have learned how to create a `ppp` object both with and without marks by importing a CSV file. Using density, we can easily visualize these points. After that, we saw how to use the quadrat test for checking CSR in a point pattern process. For checking clustering, we learned about three more functions: G, K, and L. We can test the null hypothesis of CSR using maximum absolute deviation and the sum of the squared distance between different simulated functions. We also learned how to use the `spatialkernel` package of R for spatial segregation of bivariate marked point pattern data. In the next chapter, we'll learn how to use geostatistics using different functionalities of R.

7
Spatial Analysis

This chapter will look at spatial attribute analysis and geostatistical methods, and will introduce you to the various interpolation methods for spatial data that are available in R. We'll learn how to test for spatial autocorrelation, model spatial autocorrelation, how to spatially interpolate, how to fit a **Generalized Linear Model** (**GLM**) and look at some of the most widely used geostatistical methods such as variogram and kriging.. We'll also look at how point data can be converted into raster data using interpolation methods.

We'll be covering the following topics:

- Testing spatial autocorrelation
- Modeling spatial autocorrelation
- Generalized linear models
- Geostatistics

Testing autocorrelation

In this section, we'll learn how to test spatial autocorrelation. In this test, we examine whether the assumption of the independence of observations from one another is true. In a normal distribution, we assume that observations are independent of each other, which is likely not to be true for spatial data, according to the first law of geography:

> *Everything is related to everything else, but near things are more related than distant things.*

As such, it is important to test for spatial autocorrelation when dealing with spatial data. Unlike with point pattern data, we are considering data that is picked or the location of which is fixed by the observer (for example, survey data with location information).

Preparing data

We'll be working with migration data from the Sylhet division of Bangladesh (this data is fictitious and created for the sake of demonstrating different spatial analysis examples). In the `migration.csv` CSV file, we have data for 30 *thanas* (administrative units) of Sylhet. This file has six columns, where the first column, `ID`, represents the unique ID of each area for which data is compiled; `val` represents the percentage of population in that area that have migrated; `agri` represents the percentage of agricultural area to total area; `pop` represents the population, in millions, of that area; `migration` represents the total number of people who have migrated; and `density` represents the population density for each area. We also have a shapefile, `syl_div.shp`, which has 30 polygons corresponding to the 30 *thanas* we have in the `migration.csv` file. In the `syl_div.shp` file, the unique identifier for each polygon is saved in the field named `ID_4`.

Let's import the CSV file first and have a look at the first few observations of its columns:

```
migration = read.csv("F:/Hands-on-Geospatial-Analysis-Using-R-and-
QGIS/Chapter07/Data/migration.csv")
head(migration)
```

We see that it has six columns, as expected:

```
    ID   val agri   pop migration density
1 9185 0.040 0.25 0.020       800     400
2 9186 0.020 0.29 0.050      1000     450
3 9187 0.025 0.30 0.060      1500     500
4 9188 0.050 0.26 0.060      3000     700
5 9189 0.060 0.32 0.170     10200    1200
6 9190 0.080 0.34 0.021      1680     850
```

Now, using `readOGR()` of the `spdep` library, we read in the `syl_div.shp` shapefile:

```
library(rgdal)
syl = readOGR("F:/Hands-on-Geospatial-Analysis-Using-R-and-
QGIS/Chapter07/Data", "syl_div")
```

We can also change the projection to longitude and latitude using `spTransform()` of the `sp` package in the following way:

```
library(sp)
syl = spTransform(syl, CRS("+proj=longlat +datum=WGS84"))
```

This now reprojects our data into longitude and latitude. Now we need to add the CSV file information to our `spatialPolygonsDataFrame` file, `syl`. Using `merge()` and showing the two unique identifiers for the `ID_4` shapefile and the `ID` CSV file, we can do this in the following way:

```
migration_spdf = merge(syl, migration, by.x="ID_4", by.y="ID")
class(migration_spdf)
```

We can see that the class of `migration_spdef` is `spatialPolygonsDataFrame`:

```
[1] "SpatialPolygonsDataFrame"
attr(,"package")
[1] "sp"
```

This `spatialPolygonsDataFrame` class now has all of the information from the CSV file added to each of the polygons.

Let's have a look at the percentage of migrated people for each of the polygons (*thanas*) using `spplot()`:

```
spplot(migration_spdf, zcol="val")
```

We get the following output:

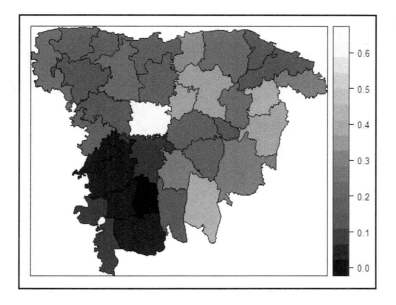

It looks as if there is a spatial pattern in migration; for example, in the southwest region, we see the lowest percentage of migration. We'll be using Moran's I index here to measure autocorrelation in the next segment.

Moran's I index for autocorrelation

Moran's I index is an index for measuring autocorrelation. To run this test, we need to get neighboring polygons for each polygon, and we'll use the `poly2nb()` function of the `spdep` package to make a list of neighbors, a type known as an `nb` object in the `spdep` package. Then, we'll need to convert this `nb` object into a `listw` object using `nb2listw()` to create a list of neighbors along with their weights. Let's first create the `nb` object, a list of neighbors for each polygon in `migration_spdef`:

```
library(spdep)
neighbor_syl = poly2nb(syl)
```

Now let's plot this `nb` object on the map to see what it has done:

```
plot(syl, col="gray")
plot(neighbor_syl, coordinates(syl), add = TRUE, col="blue")
```

Now we see that the neighborhoods are shown using connected lines:

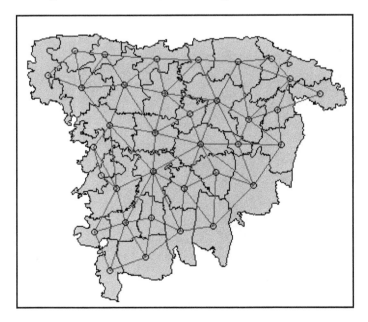

We can get the value of Moran's I index using `moran.test()` in the following way:

```
moran.test(migration_spdf$val, nb2listw(neighbor_syl))
```

This gives us the following result:

```
              Moran I test under randomisation

data:  migration_spdf$val
weights: nb2listw(neighbor_syl)

Moran I statistic standard deviate = 3.9003, p-value = 4.805e-05
alternative hypothesis: greater
sample estimates:
Moran I statistic        Expectation              Variance
     0.38835134          -0.02941176            0.01147294
```

We find that `p-value` is a very small number (much less than 0.05), and so we can reject the null hypothesis of no autocorrelation in favor of autocorrelation.

We can also use simulation to test the significance of Moran's I; we do so by doing a number of permutations of the data and assignments to polygons. We calculate Moran's I for each of these permutations and then, using this, we can get `p-value`. In R, we can use `moran.mc()` for simulation:

```
moran.mc(migration_spdf$val, nb2listw(neighbor_syl), nsim = 499)
```

Here, we have simulated 499 times, and the result is as follows:

```
              Monte-Carlo simulation of Moran I

data:  migration_spdf$val
weights: nb2listw(neighbor_syl)
number of simulations + 1: 500

statistic = 0.38835, observed rank = 500, p-value = 0.002
alternative hypothesis: greater
```

We can also see here that `p-value` is `0.002`, and so we can reject the null hypothesis of no autocorrelation.

Modeling autocorrelation

We have so far learned how to check for spatial autocorrelation, but we have not yet learned how to incorporate this into our model. We'll now learn how to do this.

Spatial autoregression

The spatial autoregression model considers the dependence of the value upon near regions and integrates that dependence into the data-generation process. We'll be working with a **Simultaneous Autoregressive (SAR)** model here. One important parameter here is lambda, which indicates the level of spatial dependence, where a positive value indicates positive correlation, a negative value indicates negative correlation, and zero means no spatial dependence.

First, we fit a SAR model without a predictor (that is, as an intercept-only model) using the spautolm() function of the spdep package. The first argument to spautolm() specifies the model and the second argument specifies the listw object, containing neighbors and weights. We make an SAR of the migration rate as a percentage as follows:

```
syl_sar = spautolm(migration_spdf$val~1, listw=nb2listw(neighbor_syl))
summary(syl_sar)
```

Here is the summary of this SAR model:

```
Call: spautolm(formula = migration_spdf$val ~ 1, listw = nb2listw(neighbor_syl))

Residuals:
     Min        1Q     Median        3Q       Max
-0.136611 -0.069291 -0.025487  0.064057  0.401192

Coefficients:
            Estimate Std. Error z value  Pr(>|z|)
(Intercept) 0.236783   0.045726  5.1784 2.239e-07

Lambda: 0.59292 LR test value: 9.6294 p-value: 0.0019148
Numerical Hessian standard error of lambda: 0.15822

Log likelihood: 25.81055
ML residual variance (sigma squared): 0.012127, (sigma: 0.11012)
Number of observations: 35
Number of parameters estimated: 3
AIC: -45.621
```

In the output, we see that the `p-value` instance corresponding to `Lambda` is 0.0019148, which is much less than 0.05, and so we can reject the null hypothesis of no spatial dependence.

Now we'll consider a SAR model with predictors. We'll check now whether the percentage of the population that migrated depends on the ratio of agricultural land to total land (here, the area of the *thana* or polygon), indicated by the `agri` column; the population density of each area being captured in the `density` column. We can do so by writing the following commands:

```
syl_sar_predictor = spautolm(val~agri+density, weights=pop,
data=migration_spdf, listw=nb2listw(neighbor_syl))
summary(syl_sar_predictor)
```

The summary of this model is as follows:

```
Call: spautolm(formula = val ~ agri + density, data = migration_spdf,
    listw = nb2listw(neighbor_syl), weights = pop)

Residuals:
      Min       1Q    Median       3Q      Max
-0.152691 -0.089436 -0.053178  0.031409  0.379258

Coefficients:
              Estimate Std. Error z value Pr(>|z|)
(Intercept) 2.0674e-01 1.0630e-01  1.9448   0.0518
agri        1.7962e-02 1.4182e-01  0.1267   0.8992
density     6.8531e-05 8.0376e-05  0.8526   0.3939

Lambda: 0.49954 LR test value: 5.2865 p-value: 0.021491
Numerical Hessian standard error of lambda: 0.1885

Log likelihood: 19.36808
ML residual variance (sigma squared): 0.0013605, (sigma: 0.036885)
Number of observations: 35
Number of parameters estimated: 5
AIC: -28.736
```

As `p-value` for all of the predictors is higher than 0.05, there is the suggestion that the proportion of agricultural land to total land and population density do not affect the migration rate. As the `p-value` of `Lambda` is less than 0.05, we can say that migration rates are spatially correlated.

Generalized linear model

The observed variable might not always be amenable to the assumptions of normal distribution and, in those cases, using a linear model is not a good idea. Instead, we can use a GLM, which allows the observed variable to have an error distribution different to the normal distribution.

Modeling count data using Poisson GLM

Suppose we have the migration rates for each of the areas we have considered so far. This is the count data, and if it can be assumed that the migration rate is constant, we can use the Poisson model to get the probability of a specified rate of migration. In a Poisson model, the mean and variance are the same.

Using the `glm()` function, we can fit a different GLM model. Here we model the rate of migration, `migration`, on `agri` and `density`, as we have done before. The code for this is as follows:

```
migration_glm = glm(migration ~ agri+density, offset = pop,
data=migration_spdf, family = poisson)
summary(migration_glm)
```

The result of this model is as follows:

```
Call:
glm(formula = migration ~ agri + density, family = poisson, data = migration_spdf,
    offset = pop)

Deviance Residuals:
    Min       1Q   Median       3Q      Max
-171.94  -127.82   -32.87    39.08   370.32

Coefficients:
             Estimate Std. Error z value Pr(>|z|)
(Intercept) 8.412e+00  6.094e-03  1380.3   <2e-16 ***
agri        1.172e+00  8.479e-03   138.2   <2e-16 ***
density     1.301e-03  4.960e-06   262.2   <2e-16 ***
---
Signif. codes:  0 '***' 0.001 '**' 0.01 '*' 0.05 '.' 0.1 ' ' 1

(Dispersion parameter for poisson family taken to be 1)

    Null deviance: 595386  on 34  degrees of freedom
Residual deviance: 516058  on 32  degrees of freedom
AIC: 516461

Number of Fisher Scoring iterations: 5
```

Well, the result is now surprising, as agri and density are both significant now. But, using a SAR model, we saw that these were insignificant. So, what happened? The answer lies in the residuals of the model. If we have correctly specified the model, the map of residuals should not show any pattern and should instead look random. So, let's map the residuals of this model on the map:

```
migration_spdf$residual = residuals(migration_glm)
spplot(migration_spdf, "residual")
```

This gives us the following map:

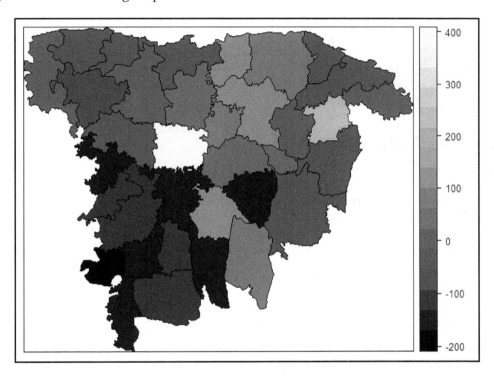

This residual map looks to have some pattern, and the residuals look like they have correlation among them. To be sure, we can further conduct Moran's I test on the residuals:

```
moran.mc(migration_spdf$residual, nb2listw(neighbor_syl), 499)
```

This gives us the following result:

```
            Monte-Carlo simulation of Moran I

data:  migration_spdf$residual
weights: nb2listw(neighbor_syl)
number of simulations + 1: 500

statistic = 0.32296, observed rank = 498, p-value = 0.004
alternative hypothesis: greater
```

As `p-value` is much lower than 0.05, we can say that the residuals are spatially correlated. This means that we need to model this data with other models, and that the Poisson model, with our specifications, is not a good fit here. We'll not discuss other ways of doing so, but we have already seen that we can model this using the spatial autoregression model.

Spatial interpolation

Spatial interpolation is a technique for predicting spatial data in a place where there is no observed data. This technique uses observed data for interpolation. Now we'll work with arbitrary meteorological data recorded in several locations of areas surrounding Dhaka, Bangladesh. We'll predict temperature at every point of the raster that covers this area. For doing so, we consider the raster of the **Digital Elevation Model** (**DEM**) and use its extent to generate a raster and plot point data on this raster. Then, using different interpolation techniques, we'll predict temperatures at every other point of the raster file.

Nearest-neighbor interpolation

We'll then convert this raster data into point data and get the nearest neighbor for each location and assign values to an unobserved point; the values will be those of the nearest neighboring points. This is called nearest-neighbor interpolation.

Let's load the DEM file first:

```
library(raster)
dem = raster("F:/Hands-on-Geospatial-Analysis-Using-R-and-QGIS/Chapter
07/Data/dem.tif")
plot(dem, main="Elevation")
```

Now have a look at the coordinate reference system of this raster screenshot:

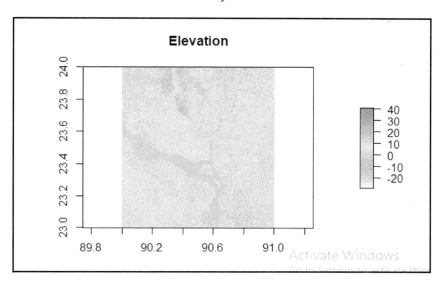

```
dem@crs
```

We get the following result:

```
CRS arguments:
 +proj=utm +zone=47 +datum=WGS84 +units=m +no_defs
+ellps=WGS84 +towgs84=0,0,0
```

We'll aggregate 4 x 4 cells to one cell for computational purpose (to get more speed):

```
dem = aggregate(dem, 4)
```

Load the CSV file with temperature data:

```
values = read.csv("F:/Hands-on-Geospatial-Analysis-Using-R-and-
QGIS/Chapter07/Data/temp.csv")
```

Now convert this DataFrame into SpatialPointsDataFrame:

```
coordinates(values) = ~ longitude + latitude
```

We need to set the coordinate reference system of this `SpatialPointsDataFrame` using CRS:

```
proj4string(values) = CRS("+proj=longlat +datum=WGS84")
```

As we need to use the raster extent for spatial interpolation, we need to transform the CRS of `SpatialPointsDataFrame` to the CRS of the raster file:

```
temp = spTransform(values, CRS(proj4string(dem)))
```

Now we convert the raster cells into points, calculate the distance between each point in the raster, and calculate the nearest point indices:

```
# Convert raster to points
raster = rasterToPoints(dem, spatial = TRUE)
library(rgeos)
# Calculate distance matrix
distance = gDistance(temp, raster, byid = TRUE)
# Calculate nearest point index
nearest_index = apply(dist, 1, which.min)
```

Now, in the attribute table of the raster, we assign new predicted temperature values that are the nearest neighbors to those points in the raster. We, again, then convert into raster data:

```
raster$value = values$value[nearest_index]
raster = rasterize(raster, dem, "value")
plot(raster)
```

We get the following result:

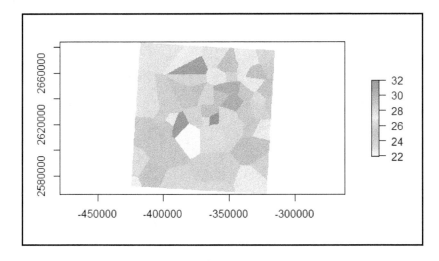

Inverse distance weighting

In **Inverse Distance Weighting (IDW)**, the predicted value at a point is a weighted average of the values measured at different points. Here, we create a model without independent variables with the following command:

```
install.packages("gstat")
library(gstat)
g = gstat(formula = value ~ 1, data = temp)
```

By using `gstat()`, we create a `gstat` object that can now be utilized to make predictions of temperature using `interpolate()` in the following way:

```
prediction = interpolate(dem, g)
```

Now plot the predicted raster along with the observed data:

```
plot(prediction)
plot(temp, add = TRUE, pch = 18, cex = 0.5)
```

We see that we now have a much smoother prediction of temperature over a continuous area:

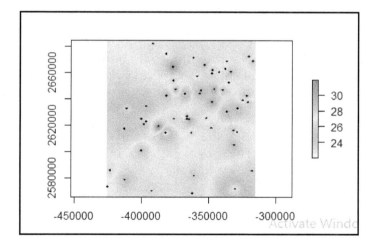

Geostatistics

In geostatistics, observed variables are random, with the assumption that we observe them as outcomes of random processes. Usually, the data is observed in discrete locations of a spatially continuous regions, and we are interested in the prediction of values in non-observed locations in the given region. For example, soil salinity is measured in certain points in an area, and we are interested in predicting or interpolating soil salinity in every point under this area. Now, soil salinity in a point could depend on multiple factors: it could be due to its proximity to the saline-affected area, its spatial autocorrelation, the climate, or a random process. Other examples of geostatistical data include rainfall data, air quality, and measurements of chemical components at multiple locations in an area.

Some important concepts

Covariance is a measure of dispersion. An analogous measure for geostatistical data is a **covariogram**, which is the same as covariance with the only difference being that observations are spatially indexed and covariance between points is measured at a fixed separation distance. Similar to the case with correlation, we have a **correlogram** for spatial data where, again, observations are spatially indexed and we take correlation between pairs of points at a given separation distance. Now, if we plot a correlogram for all possible separation distances, or lags, we get a correlogram plot.

We need second-order stationarity to explain local variation using covariograms. **Second-order stationarity** conditions stipulate that the expected value of the variable is independent of the points and that the covaroigram for any distance is independent of the points.

We'll work with the `meuse` dataset from the `sp` package. Let's load this dataset and use its coordinates to turn it into a `SpatialPointsDataFrame` instance:

```
library(sp)
data(meuse)
coordinates(meuse) = c("x", "y")
head(meuse@data)
```

We now see the first five rows of this `meuse` dataset as follows:

```
  cadmium copper lead zinc  elev       dist   om ffreq soil lime landuse dist.m
1    11.7     85  299 1022 7.909 0.00135803 13.6     1    1    1      Ah     50
2     8.6     81  277 1141 6.983 0.01222430 14.0     1    1    1      Ah     30
3     6.5     68  199  640 7.800 0.10302900 13.0     1    1    1      Ah    150
4     2.6     81  116  257 7.655 0.19009400  8.0     1    2    0      Ga    270
5     2.8     48  117  269 7.480 0.27709000  8.7     1    2    0      Ah    380
6     3.0     61  137  281 7.791 0.36406700  7.8     1    2    0      Ga    470
```

We'll only work with data in the `zinc` column, and so now we subset this data from the `meuse` dataset:

```
meuse_coord = meuse@coords
zinc = meuse@data[, 4]
# cbind this two
meuse_zinc = cbind(meuse_coord, zinc)
head(meuse_zinc)
```

This gives us the following dataset, as we expected:

```
       x      y zinc
1 181072 333611 1022
2 181025 333558 1141
3 181165 333537  640
4 181298 333484  257
5 181307 333330  269
6 181390 333260  281
```

Now let's map the `zinc` column of the `meuse` dataset to have an understanding of how it looks:

```
spplot(meuse, "zinc")
```

This gives us the following plot:

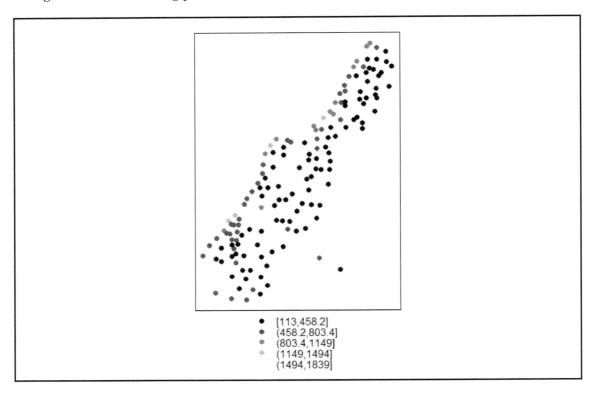

Many values of this `zinc` column are concentrated in the lower range and, if we use `do.log=TRUE`, these values pop up in the plot:

```
spplot(meuse, "zinc", do.log=TRUE)
```

We get the following result:

We now see that the values are now segregated in a better way for visualization, as many values fall between 113 and 197.4 and 197.4 and 344.9 (see the preceding plot).

We'll use the `gstat` and `geoR` packages to demonstrate how to perform different geostatistical analyses. Now we'll convert the `meuse_zinc` DataFrame into a class of `geodata` defined in the `geoR` package using the `as.geodata()` function:

```
meuse_zinc = as.geodata(meuse_zinc)
class(meuse_zinc)
```

```
[1] "geodata"
```

We see that the type of data is now `geodata`. Now we can apply various functions in the `geoR` package. We'll, in fact, use both the `meuse` dataset's `SpatialPolygonsDataFrame`, which is to be used with the `gstat` package; and the `meuse_zinc` dataset, which is to be used with the `geoR` package.

Let's also plot these points depending on which quantiles they belong to, indicated by size and color, using the `points.geodata()` function of the `geoR` package as follows:

```
points.geodata(meuse_zinc, xlab="X", ylab="Y", pt.divide="quintile")
```

This gives us the following plot:

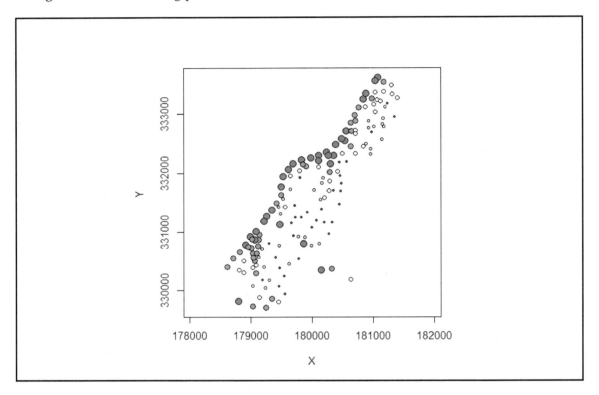

Variograms

Semivariance is the squared difference between two points divided by two. A **variogram** cloud plots all of the squared differences of all pairs of points as a function of separation distance. A variogram is the expected squared difference of all pairs of points for a separation distance.

The **nugget** is the value of the semivariance at the starting location, which reflects the measurement error. The **sill** is where the semivariogram reaches its maximum height. The **range** is the distance beyond which the variogram value doesn't change much. A partial sill is defined as the difference between the sill and the nugget.

There are different variogram models, including pure nugget effects, linear models, spherical models, Gaussian models, and parabolic models. Depending on the plot of the variogram, we can decide which model to fit.

We now plot the variogram cloud and variogram using the `geoR` package and the `variog()` function:

```
#Set the chart to one row, two columns
par(mfrow=c(1,2))
# Variogram cloud
plot(variog(meuse_zinc,option="cloud"),main="Variogram Cloud")
#Variogram
plot(variog(meuse_zinc),main="Binned Variogram")
# draw a line
lines(variog(meuse_zinc))
# set the default chart setting to 1 row, 1 column
par(mfrow=c(1,1))
```

This gives us the following plot:

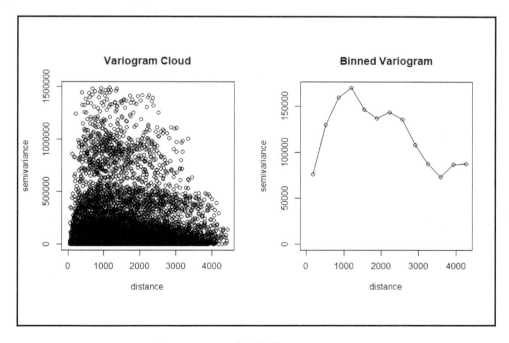

We see that semivariance reaches its maximum at about a distance of 1,500 units, and then it comes down. From the preceding variogram, it looks like the value of the nugget is 750,000, and the value of the sill is 180,000 as, at this point, semivariance reaches the maximum height.

Now, we'll use the `variogram()` function of the `gstat` package to plot the variogram:

```
library(gstat)
plot(variogram(zinc ~ 1, meuse))
```

We get the following result:

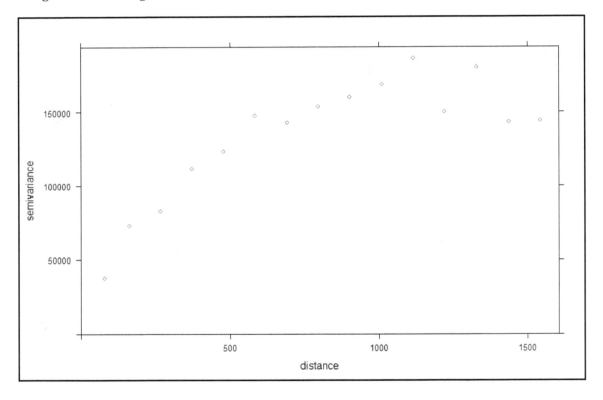

Note that here, a distance of up to 1,500 is considered instead of the distance 4,000+ considered in the variogram fitted by `variog()` in the `geoR` package. The reason for this is that, as we go further away from a point, we see a greater variation in semivariance value, which actually doesn't add much information, and so `gstat` uses a cutoff value.

Let's have a look at the `summary` of this dataset:

```
summary(meuse_zinc)
```

We get the following output:

```
Number of data points: 155

Coordinates summary
         x      y
min 178605 329714
max 181390 333611

Distance summary
       min        max
  43.93177 4440.76435

Data summary
       Min.    1st Qu.    Median      Mean   3rd Qu.      Max.
   113.0000   198.0000   326.0000  469.7161  674.5000 1839.0000
```

We see that the minimum distance is 43.93 and the maximum distance is 444.76. Now we'll plot between 40 and 1,600 in the following way:

```
# Variogram
model = variog(meuse_zinc,uvec=seq(40,1600,l=15),bin.cloud=T)
plot(model,main="Variogram Cloud", bin.cloud=T)
```

We get the following output:

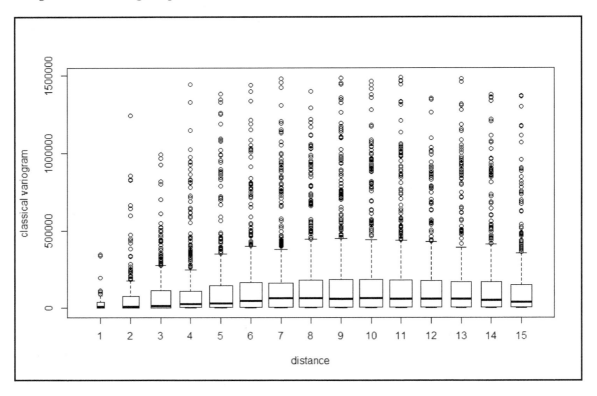

We see that there are 15 bins, clouds, and box plots together.

Now, let's use the `variofit()` function of the `geoR` package to automatically fit a line to it:

```
model = variog(meuse_zinc)
model_fit = variofit(model)
model_fit
```

We get the following result:

```
variog: computing omnidirectional variogram
variofit: covariance model used is matern
variofit: weights used: npairs
variofit: minimisation function used: optim
initial values not provided - running the default searchvariofit: searching for best initial value ... selected
values:
              sigmasq     phi   tausq kappa
initial.value "127667.62" "0"   "0"   "0.5"
status        "est"       "est" "est" "fix"
loss value: 10312523430301.1
variofit: model parameters estimated by WLS (weighted least squares):
covariance model is: matern with fixed kappa = 0.5 (exponential)
parameter estimates:
     tausq    sigmasq        phi
  3537.623 131205.241      0.000
Practical Range with cor=0.05 for asymptotic range: 0.0001159668

variofit: minimised weighted sum of squares = 9.715118e+12
```

Here `sigmasq` is the sill value, and we can see that the value of this is `131205.241`.

Using `vgm()` from the `gstat` package, we can also model a variogram. A list of the models that we can fit to the sample variogram can be listed as follows:

```
vgm()
```

We get the following output:

```
   short                                     long
1    Nug                              Nug (nugget)
2    Exp                         Exp (exponential)
3    Sph                           Sph (spherical)
4    Gau                            Gau (gaussian)
5    Exc        Exclass (Exponential class/stable)
6    Mat                              Mat (Matern)
7    Ste Mat (Matern, M. Stein's parameterization)
8    Cir                             Cir (circular)
9    Lin                              Lin (linear)
10   Bes                              Bes (bessel)
11   Pen                       Pen (pentaspherical)
12   Per                            Per (periodic)
13   Wav                               Wav (wave)
14   Hol                               Hol (hole)
15   Log                         Log (logarithmic)
16   Pow                              Pow (power)
17   Spl                              Spl (spline)
18   Leg                            Leg (Legendre)
19   Err                  Err (Measurement error)
20   Int                          Int (Intercept)
```

Now we'll fit the variogram parameters of a spherical model to the sample variogram:

```
model2 = variogram(zinc~1, meuse)
fit.variogram(model2, vgm("Sph"))
```

We get the following result:

```
   model    psill     range
1    Nug  24802.7    0.0000
2    Sph 134746.5 831.0127
```

`vgm()` can also take a set of models and then return the best model, as follows:

```
model2.fit = fit.variogram(model2, vgm(c("Mat", "Exp", "Gau", "Sph")))
model2.fit
```

We get the following output:

```
   model    psill    range kappa
1    Nug   9486.4   0.0000   0.0
2    Mat 163285.3 381.7076   0.5
```

We can see that it returns the `Mat` model as the model of best fit.

Kriging

Kriging is an interpolation process in which interpolated values are due to a Gaussian process. In kriging, a variogram is used to interpolate geostatistical data. Kriging gives us the best, linear, unbiased prediction.

With the use of observed data and a variogram, kriging computes estimates and uncertainties at unobserved points. There are different types of kriging; we'll discuss some of them briefly now. Ordinary kriging assumes that the mean is constant. Simple kriging assumes that the generalized least squares estimate of the trend coefficients is known. Universal kriging has a local trend component.

Now we'll use some of these kriging methods. We'll use the `meuse.grid` dataset, which contains coordinates of points on a regular grid. We convert it into `SpatialPixelsDataFrame`:

```
data(meuse.grid)
coordinates(meuse.grid) = c("x", "y")
meuse.grid = as(meuse.grid, "SpatialPixelsDataFrame")
```

Now we fit a simple kriging as follows:

```
krig_simple = krige(zinc ~ 1, meuse, meuse.grid, model2.fit, beta = 10)
```

Let's have a look at the names of different contents of `krig_sample`:

```
names(krig_simple)
```

We get the following result:

```
[1] "var1.pred" "var1.var"
```

Let's plot the predicted values contained in `var1.pred`:

```
spplot(krig_simple, "var1.pred")
```

We get the following plot:

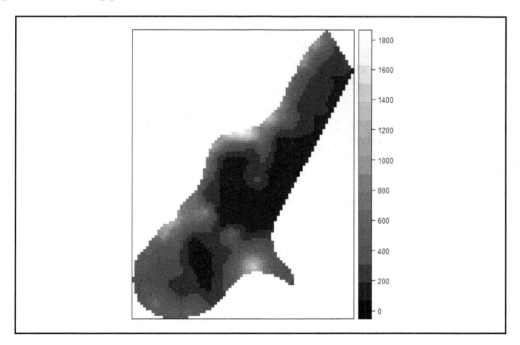

We can also compute exceedance probability; for example, we can compute the probability of a value of `zinc` being above 1,200:

```
krig_simple$exceedanceProb = 1 - pnorm(1200, mean = krig_simple$var1.pred,
sd = sqrt(krig_simple$var1.var))
```

Now plot the exceedance probability as follows:

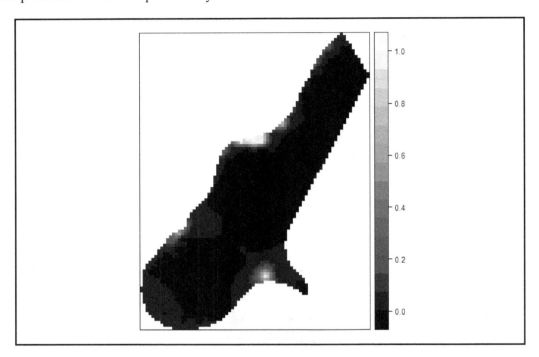

It looks like, for most of the area, the exceedance probability is very low.

Checking residuals

We can check how well our kriging model is doing in modeling the data and in predicting. We can use cross-validation, which divides data into two parts—training and validation—where we fit the variogram model to the training set and the kriging model on the validation set. We then compare prediction on the validation set with the measured value on the training set.

Let's create training data with 70% of the `meuse` data and create validation data with the remaining 30%:

```
no_rows = dim(meuse@data)[1]
sample_size = floor(0.75 * no_rows)
train_no = sample(seq_len(no_rows), size = sample_size)
train = meuse[train_no, ]
validation = meuse[-train_no, ]
```

Now, fit a variogram to the training data:

```
# In vgm, first argument is partial sill, second is model, third is range
train_fit = fit.variogram(variogram(zinc~1, train), vgm(134746.5, "Mat",
1200, 1))
```

Get a kriging prediction for the validation set as follows:

```
validation_pred = krige(zinc ~ 1, train, validation, train_fit)
```

Now, we'll estimate the R2:

```
kriging_residual = validation$zinc - validation_pred$var1.pred
mean_residual = validation$zinc - mean(validation$zinc)
# R2 computation
1 - sum(kriging_residual^2)/sum(mean_residual^2)
```

We get the following output:

```
[1] 0.4787689
```

We see that our R2 is only 0.47, which is a poor fit. That means we have to look for other ways to improve the model fitting and prediction.

Summary

In this chapter, we learned how to test for autocorrelation in spatial data using Moran's I index. We also learned how to model autocorrelation using a SAR model. This was followed by modeling count data using a Poisson GLM. After that, we learned the basics of geostatistics, with variograms and kriging, in particular. Then, we learned how to understand the degree of spatial dependence in a random field using variograms. We learned about the different ways of plotting and modeling spatial dependence using the `geoR` and `gstat` packages. We then learned how to predict or interpolate data from covariograms using a method called kriging, and we learned how to compute exceedance probability and looked at checking the residuals of a prediction.

In the upcoming chapters, we'll learn how to automate different spatial processing and analysis tasks. We'll then learn how to use the power of QGIS inside R and vice versa. Finally, we'll learn how to fit machine learning models to create a landslide susceptibility model and corresponding map using real data from Bangladesh.

GRASS, Graphical Modelers, and Web Mapping

8

In this chapter, we will look at some of the functionalities of the free and open source software **Geographic Resources Analysis Support System (GRASS)** GIS, which can be accessed using QGIS. Besides this, we will also look at how graphical modelers can be used to automate repetitive tasks in QGIS. We will also learn how to use create interactive maps in both QGIS and R that can be uploaded to the web.

This chapter covers the following:

- GRASS GIS
- Graphical modeler for automating analysis
- Web mapping

GRASS GIS

GRASS is a free and open source software that is installed by default along with QGIS Desktop. We can use GRASS both as a stand-alone software or from within QGIS.

Basics of GRASS GIS

Storage of data in GRASS is different from QGIS, and we need to be familiar with the basics before we can start working on the different functionalities provided in GRASS. We need to understand the following terminology before going deeper into GRASS:

- Database
- Location
- Mapset

Database

Before we start working on a GRASS GIS project, we need to connect to a **database**. This database is a folder containing a location and a mapset, which are again subdirectories under the database folder.

Location

A **location** contains the coordinate system or map projection of all the spatial data that will be used in a project, and it also has information on map projection and geographical boundaries. A subset of a location bound to different spatial operations using a rectangular box is called a **region**.

Mapset

A **mapset** is a required subdirectory under a location that contains spatial data. Using a mapset, a project can be subdivided into different regions. There are two types of mapset:

- **Permanent**: This type of mapset contains read-only data.
- **Owner**: This contains project-specific spatial data.

Creating a mapset

In creating a mapset, the first step is to create a connection to a database. But, first, we will open **QGIS Desktop with GRASS** instead of **QGIS Desktop**, which is what we have used so far:

Now **QGIS Desktop with GRASS** will open, which looks the same as **QGIS Desktop**. We need to install the plugin for GRASS by clicking on **Manage and Install Plugins** under **Plugins**:

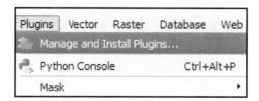

The **Plugins** window will now appear. Select **GRASS 7** and then click **Install Plugin**. After the plugin is installed, click **Close** to start working with this plugin:

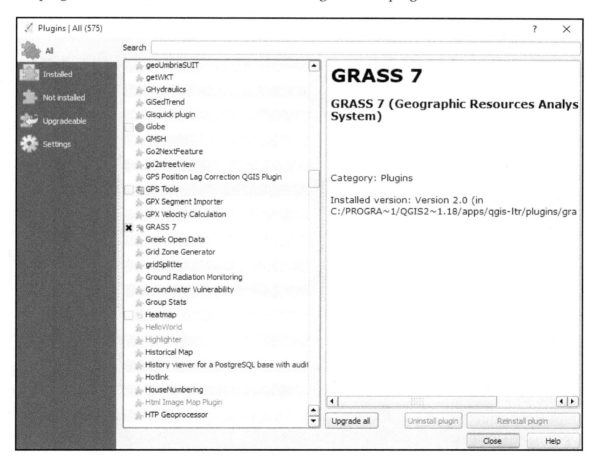

GRASS comes under the **Plugins** menu. Click **GRASS** under **Plugins** and then click on **New Mapset**:

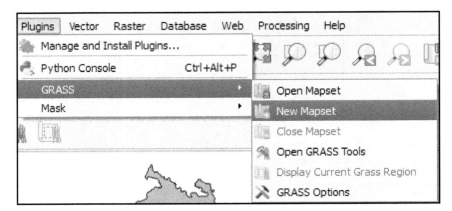

This prompts us with a new window that asks for **Database directory**. This is the directory inside which we need to keep all of the spatial data that is to be used for GRASS operations:

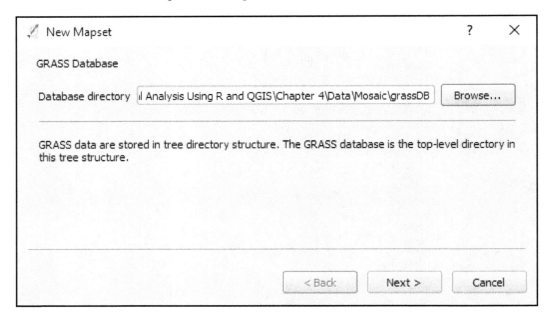

Browse to the DB_GRASS directory under the Data folder of Chapter 8 (if it is not created, create a new folder called DB_GRASS in this location):

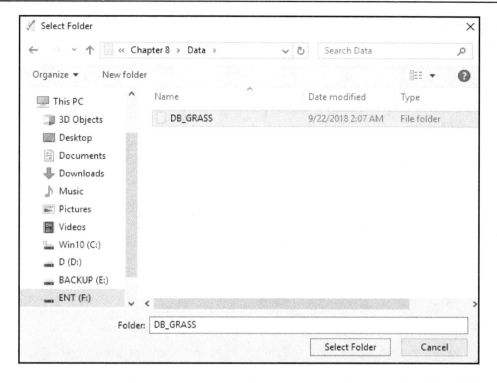

Now, we are prompted with a new window, **New Mapset**, asking for the **GRASS Location**. Name the location BD and click **Next**, as shown in the screenshot here:

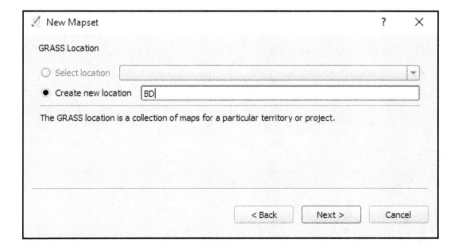

We now see options for defining coordinate reference systems. Write `4326` in **Filter** and select **WGS 84** corresponding to **EPSG:4326**, and then click **OK**:

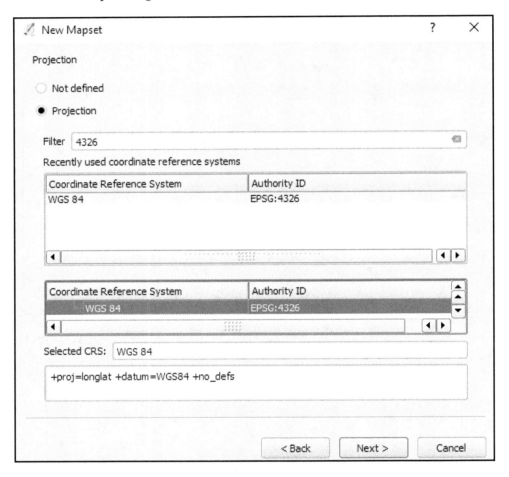

Now, we need to select the region for this mapset. Choose `Bangladesh` from the dropdown, as we will be working with data from Bangladesh, and then click **Next**:

Now, we need to give the mapset a name; in this case, write `mapset1` (or any other name of your choosing) and click **Next**:

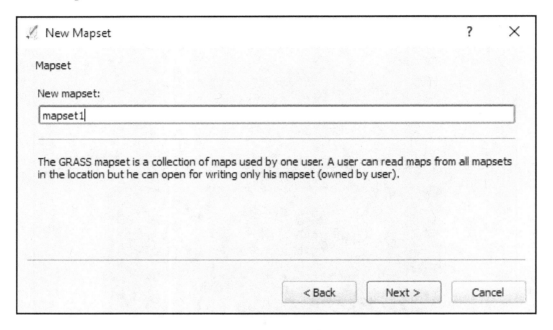

We will now see a new window similar to the following. Click on **Finish** and then click **OK**. When successful, a final success message will pop up:

Importing vector data in GRASS

We can add a vector layer as before by clicking on **Layer** in the top menu, then on **Add Layer**, and then on **Add Vector Layer**, before selecting `BGD_adm.shp`:

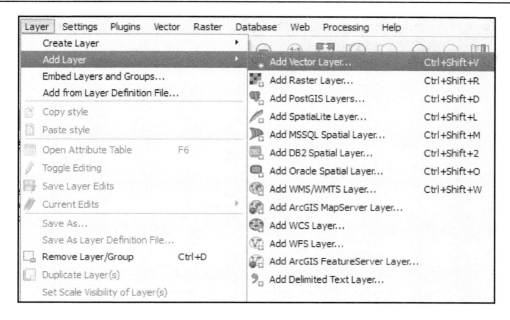

The shapefile will be loaded:

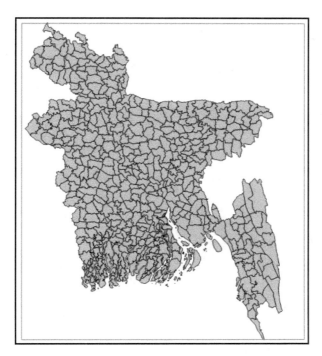

The red box surrounding the map confirms that this file is shown using the facilities of GRASS GIS. Now, to import it into GRASS GIS, we need to write `v.in.ogr.qgis` in **Filter** under **GRASS Tools**:

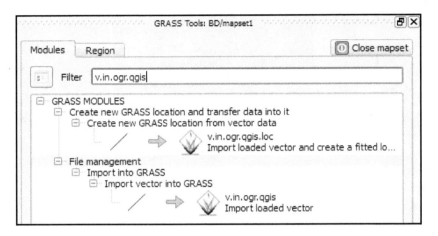

We now double-click on `v.in.ogr.qgis`, which prompts us with another window. Give a new name for the output vector map, `BGD_GRASS`, and then click **Run**:

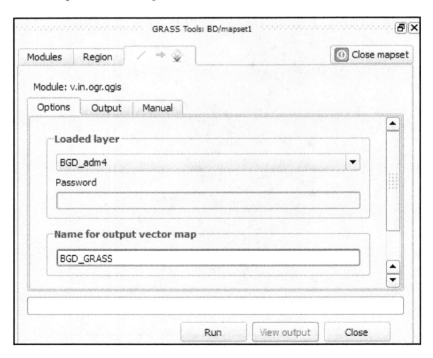

Now we will see a new message, **Successfully finished**, which means that we have successfully imported the vector data into QGIS:

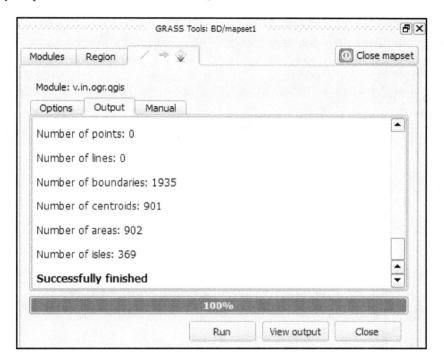

Importing raster data in GRASS

We will now import raster data into GRASS QGIS. But, first, we add the LC08_L1TP_137043_20180410_20180417_01_T1_B2.TIF raster layer into QGIS. We have to ensure that the raster data is in the WGS 84 coordinate reference system, as we have defined WGS84 as the coordinate reference system for our new mapset. We need to reproject the raster into this WGS 84. To do this, we click on **Raster** in the top menu and then click on **Projections**, followed by **Warp (Reproject)**, as shown in the following screenshot:

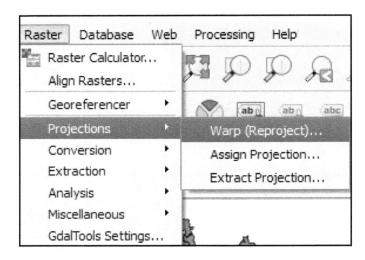

Now, select any output location and write EPSG:4326 in the **Target SRS** area. Select **Near** as **Resampling method** and select **Load into canvas when finished**, and then click **OK**:

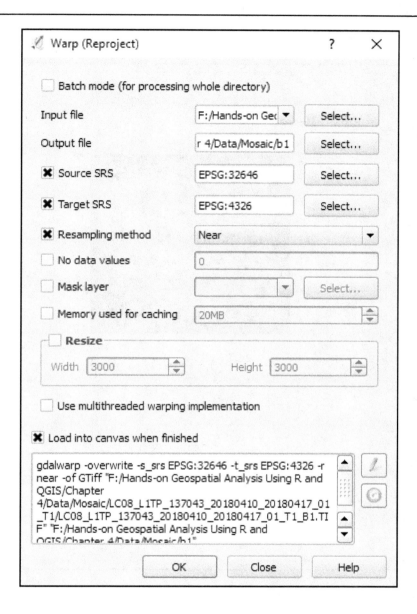

Now, we see the raster file with a red, rectangular, hollow box around it, as follows:

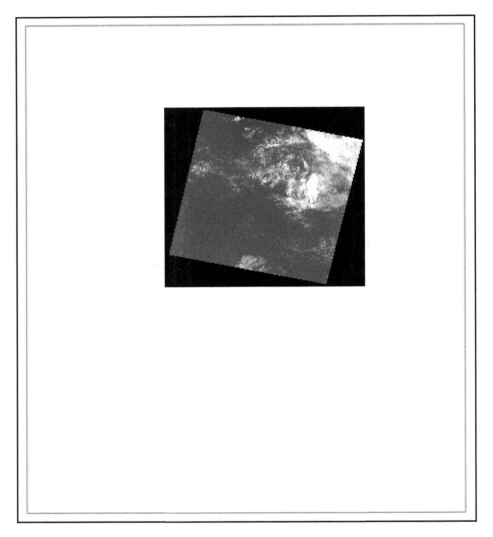

Now, to import raster data into GRASS, write `r.in.gdal.qgis` in **Filter** inside **GRASS Tools**. Click on that:

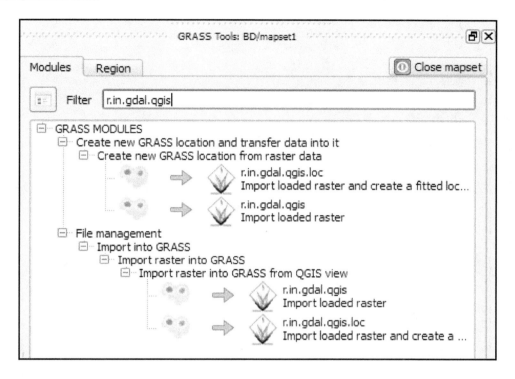

We now give a new name for our output raster map; we name it b1_GRASS here. This will now save b1_GRASS in our mapset. Finally, click **Run**, and then after it has finished running, click **Close**:

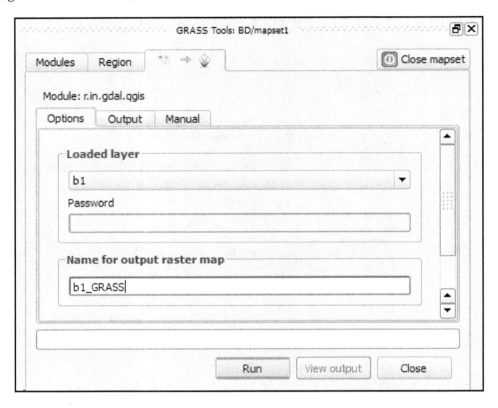

Similarly, do this for Band 4, Band 5, and Band 7, and save them as b4_GRASS, b5_GRASS, and b7_GRASS, respectively.

False color composite in GRASS

We will create a false color composite in GRASS QGIS. First, we click on **GRASS** and then on **Open Mapset**, as shown in following screenshot:

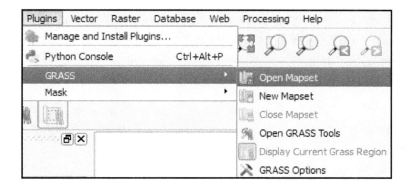

Now, we select the DB_GRASS database that we created before and select **mapset1**, as follows:

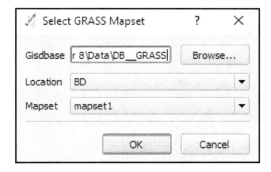

Within **GRASS Tools**, in **Filter**, type r.composite. Click on **r.composite**:

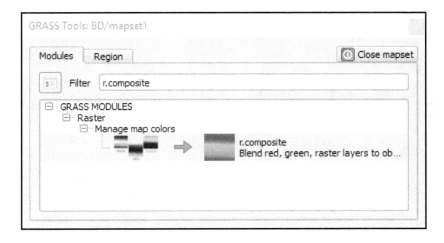

Select `b4_GRASS` for red, `b5_GRASS` for green, and `b7_GRASS` for blue. Write `composite` in the **Name for output raster map** section and then click **Run**:

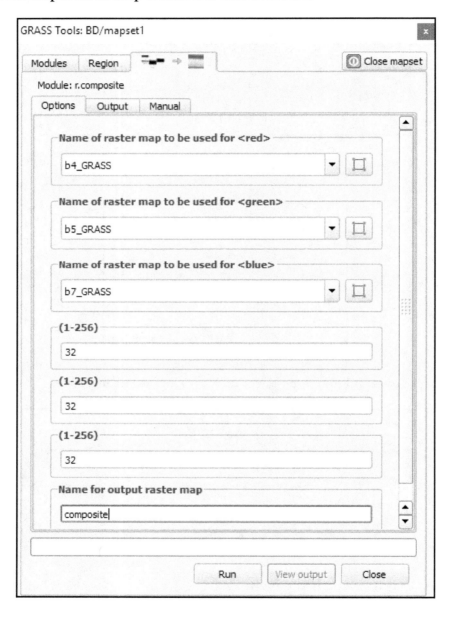

If everything goes well, we will see something like this:

The composite map will look like the following raster screenshot:

Graphical modeler

The graphical modeler provides a wonderful functionality in QGIS that helps us to automate repetitive tasks in QGIS. Suppose we want to clip a raster file according to a shapefile. The steps for this clipping for different sets of rasters and vector files would require us to repeat the same number of steps again and again for different combinations. We can reduce our workload by using a graphical modeler.

Here, we will clip the LC08_L1TP_137043_20180410_20180417_01_T1_B2.TIF raster file by the gazipur shapefile. Let's see how they look by loading them into QGIS:

The **Graphical Modeler** can be accessed under **Processing**:

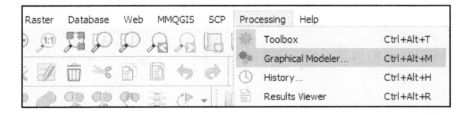

Now **Processing Modeler** will open. We see that there are two tabs: **Inputs** and **Algorithms**. Using the **Inputs** tab, we define the inputs for our model; and using **Algorithms**, we state what to do with those inputs:

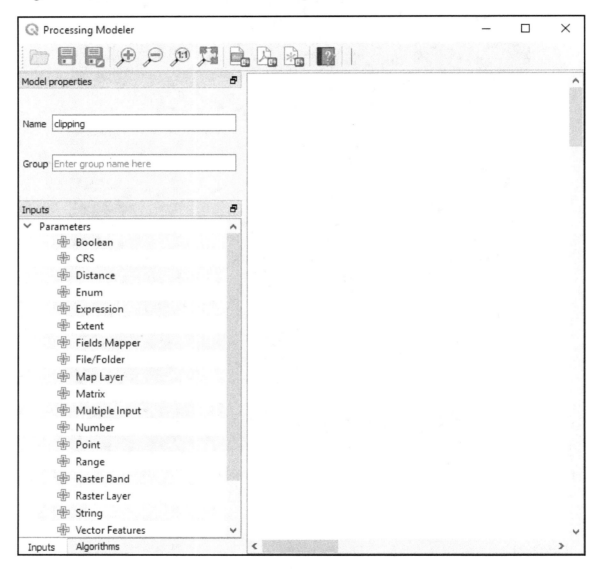

First, double-click on **Raster Layer** under the **Inputs** tab. Give a name for the parameter; here, write `landsat`. Click on **Mandatory** and then click **OK**:

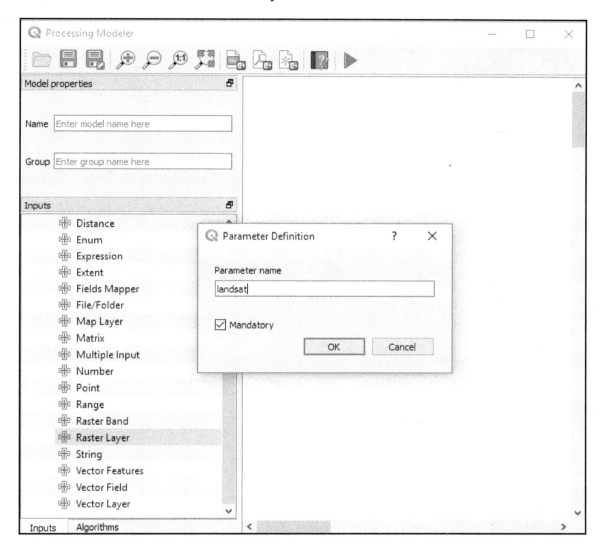

Now double-click on **Vector Layer** under the **Inputs** tab. For **Parameter name**, we write district, but you could provide any name that you feel is appropriate. Select the **Geometry type** as **Polygon** and click on **Mandatory**, then click **OK**:

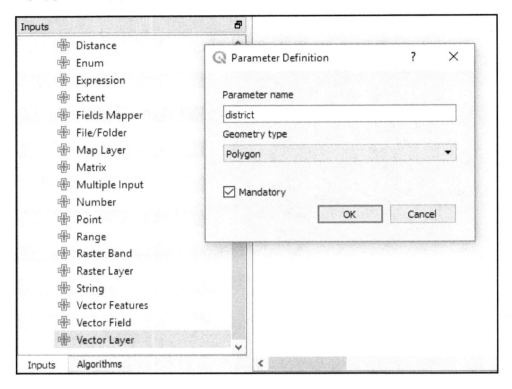

Now, the right side of the window will look similar to what we see here:

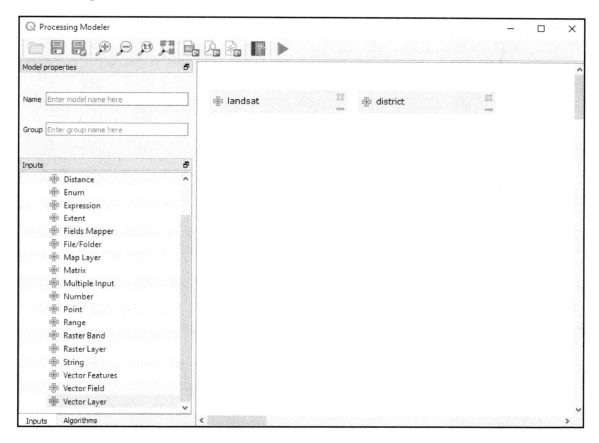

Now click **GDAL**, then **Raster extraction**, and then double-click on **Clip raster by mask layer**. Select **landsat** as **Input layer** and **district** as **Mask layer**. Give the clipped image a name **clipped** is what we have provided. Click **OK**:

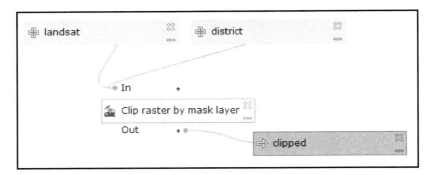

Click on the **Play** button in the top menu to run this. This will prompt us with yet another window. Under **landsat**, select the Band 2 image, and under **district**, select **gazipur**:

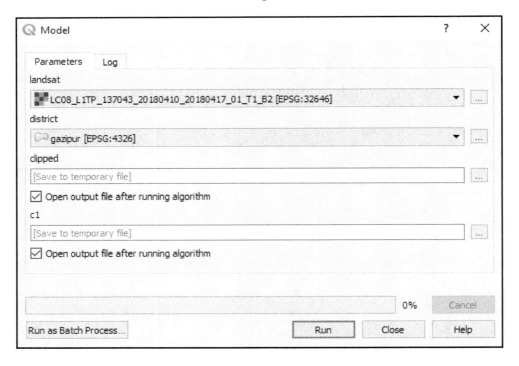

Click **Run** and now we will get a clipped raster:

To remove the black region, right-click the **clipped** layer and click on **Properties**. Click on **Transparency** and write 0 in the **Additional no data value** box:

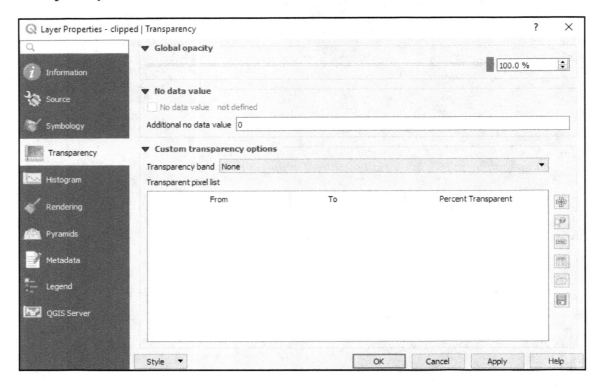

Now, we see that the black region is removed from the raster file:

Web mapping

Web mapping can be useful to us; it allows our work to be exported to a web page. We will learn now how to export maps to web maps.

Web mapping in QGIS

We need to install the `qgis2web` plugin first to be able to export to a web map; particularly an OpenLayers or Leaflet map.

Here we will work with `BGD_adm3_data.shp`, and we will colorize polygons according to the different values of the `value_Sh_2` attribute. To do so, we need to first right-click on the **BGD_adm3_data** layer and then click **Properties**. Then we need to click on **Symbology**, select **Categorized**, select `value_Sh_2` as **Column**, click on **Classify**, and then click **OK**:

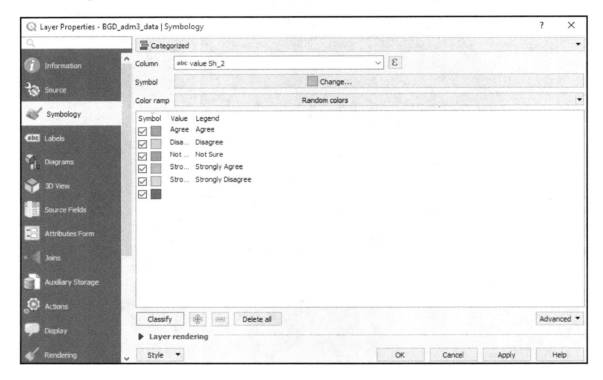

Now, we will get the following map, or something similar, at least:

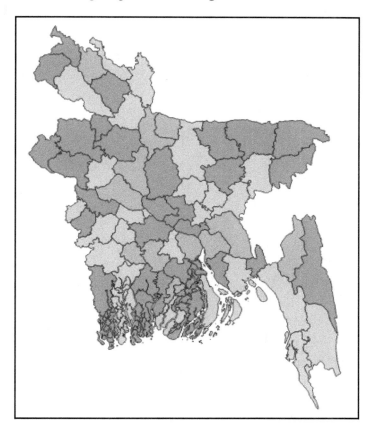

Now, to create a web map, click **Web** in the top menu, then click **qgis2web**, and then click **Create web map**:

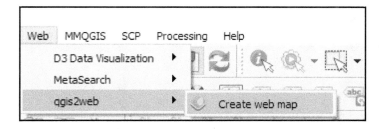

Now, we get an **Export to web map** window. Under the **Layers and Groups** tab, select **header level** only for the unique polygon ID NAME_3:

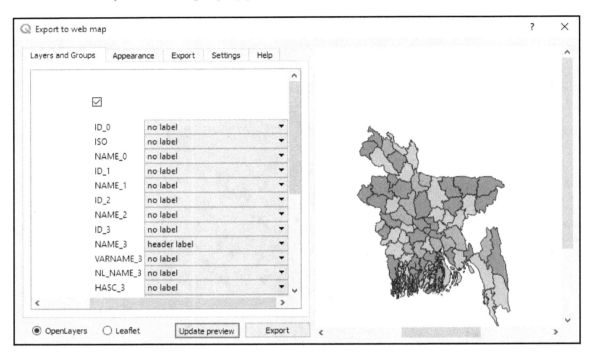

Now, under the **Export** tab, select the folder where we want the web map to be created and then click on **Export**:

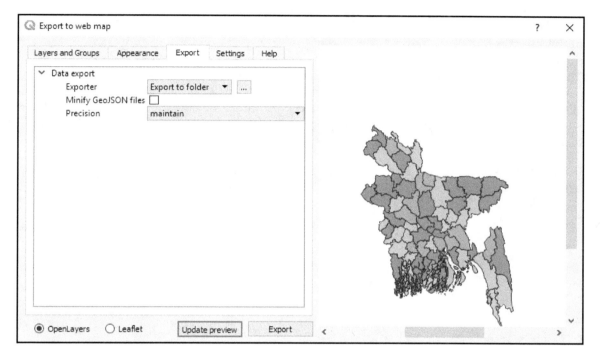

Now, we will get a web map of it in the folder we specified a few steps back.

Summary

This chapter introduced you to a slightly advanced concept in QGIS. We focused on using some functionalities of GRASS. GRASS has many other functions that we haven't covered in this chapter and, after getting knowledge from this book on the basics of GRASS, we expect that you will have some comfort in exploring other functionalities independently. This chapter only touched upon some of the key concepts in GRASS GIS that are compatible with QGIS, such as learning how to use a mapset and how to put data inside it. We also learned how to use Processing Toolbox – or, more specifically, a graphical modeler – to automate different spatial tasks. Finally, we learned how to use the `gis2web` plugin for web mapping directly from QGIS.

In the next chapter, we will familiarize you with the concept of raster image classification using the SCP of QGIS.

9
Classification of Remote Sensing Images

This chapter will familiarize you with the concept of raster image classification using the **Semi-Automatic Classification Plugin** (**SCP**) of QGIS. There are two major types of classification: supervised and unsupervised. We will cover supervised classification here, which uses training data to classify the whole dataset or a raster image. Learning about this will help to classify land cover according to different requirements or characteristics.

The following topics will be covered in this chapter:

- Definition of supervised classification
- Supervised classification in QGIS
- Creating validation data in QGIS

Classification of raster data

Classification refers to classifying data to different categories; in the case of remote-sensing literature, this refers to classifying different land cover types, generally. This can be done in two ways: supervised and unsupervised. **Supervised classification** refers to a situation where we create a training area and generate a signature area from the training area, and then use that to classify a raster. In **unsupervised classification**, the user determines into how many clusters they want to divide the data beforehand, and then the image is classified into that number of clusters; finally, the user then identifies the relevant land classes for the clusters.

Supervised classification

In supervised classification, training data is used for classification. This training data is made in such a way that it is representative of the classes or land cover types we want to classify. An unclassified image is classified using the spectral signature of the pixels in the training data or area. There are three main supervised classification algorithms that are used in QGIS: minimum distance, **maximum likelihood (ML)**, and **spectral angle mapper (SAM)**. For minimum distance, a pixel is assigned to a class that has a lower Euclidean distance to mean vector of a class than all other classes. In ML, each pixel is assigned to the class that has the highest probability. The SAM algorithm works by computing the angle between the mean vector of the class and the unclassified raster data, and the class for which the angle is the smallest is assigned to be the class of the unclassified pixels.

Supervised classification in QGIS

We will classify landsat 8 images looking at different land use patterns. We will use Band 3 (red), Band 4 (near infrared), and Band 5, or **shortwave infrared (SWIR)**, here. We use the band combination 5, 4, and 3 for a **color infrared (CIR)** image. In the following outputs, vegetation will pop up as red, and dead or unhealthy plants will appear in less intense red, green, pink, or tan. Sandy soil will appear as white, gray, or light tan, and water and roads will appear as shades of blue to black.

First, we load bands 3, 4, and 5 from `folder 2` under the `Mosaic` folder, which is inside the `Data` folder of `Chapter 4`:

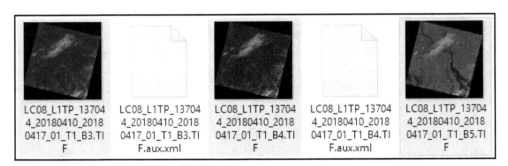

As landsat is multispectral and as we want to work with three bands, we need to make a false color composite using these three bands. To merge these three raster bands, click on **Raster**, then on **Miscellaneous**, and then on **Merge**, as follows:

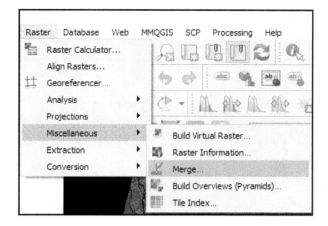

Now, we will see this window. Click on **Place each input file into a separate band** so that we can assign different bands to different colors of red, green, and blue later. We need to select all three of these rasters for **Input layers**. We do so by clicking on the **...** box to the right of the **Input** layers:

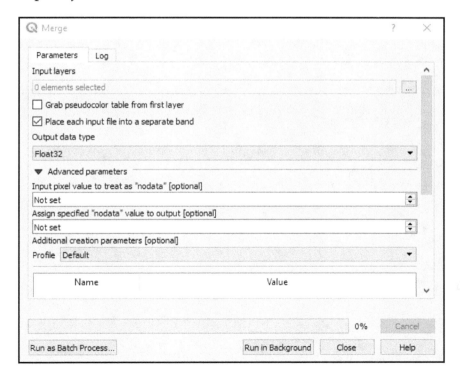

Now, tick all the required bands or rasters, or for this case, click **Select all**, and then click **OK**:

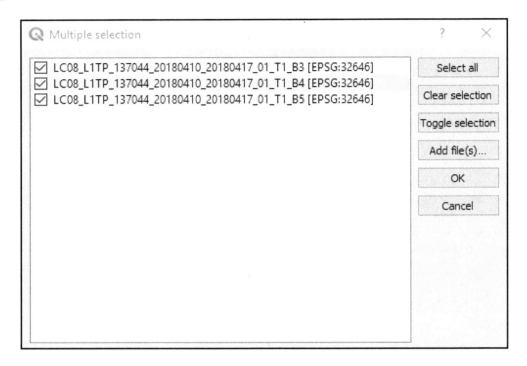

At the bottom of the **Merge** window, under **Merged**, browse to the folder where we want to save the file and give a name (here, the name we give is `merged_b3_b4_b5.tif`). Click on **Open output file after running algorithm** and then click on **Run in Background**:

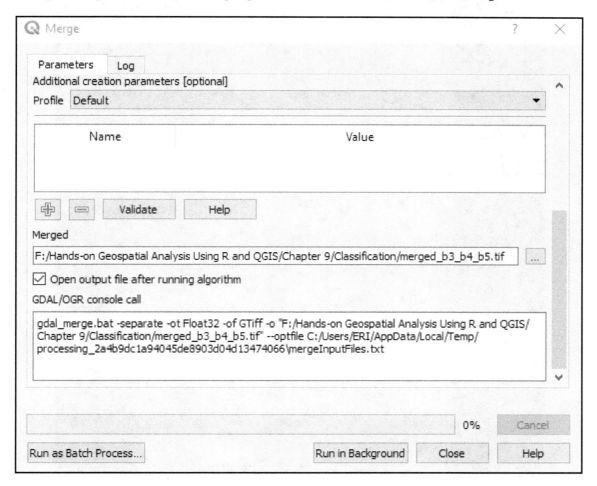

In seconds, we will see the merged TIF file. But, **RGB** (short for **red**, **green**, and **blue**) color has not been assigned as we would want:

We can do so by right-clicking on this raster and clicking on **Properties**. Click on **Symbology**, then select **Multiband color** as **Render type**. Now, assign **Band 3** to **Red band**, **Band 2** to **Green band**, and **Band 1** to **Blue band**. Click on **Apply** and then click on **OK**:

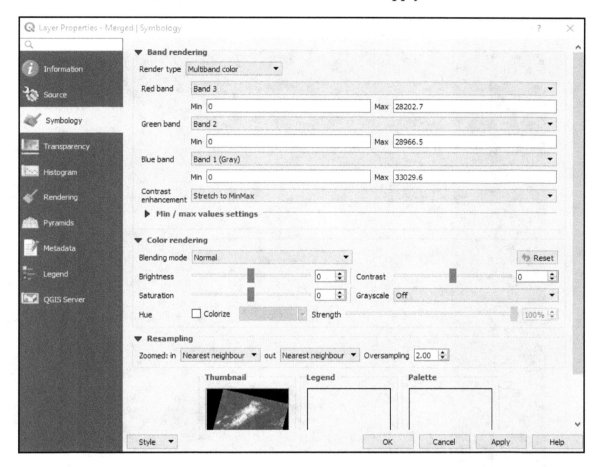

Now, we will get the following CIR image. With the naked eye, we can now differentiate land cover classification to some extent, but as we zoom into this image, it becomes clear that this classification is rather nuanced, and if we wanted to digitize this image by ourselves, the process would be very time-consuming and prone to error. This is where supervised classification comes in:

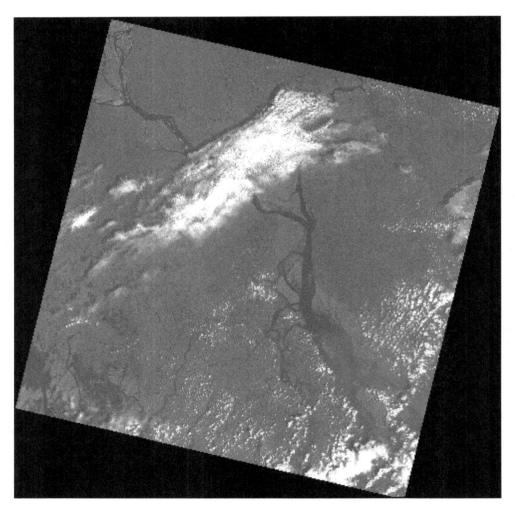

We can also use the SCP in QGIS to get a CIR image. We need to install the SCP from **Plugins** and then, under **SCP**, click on **Band set**:

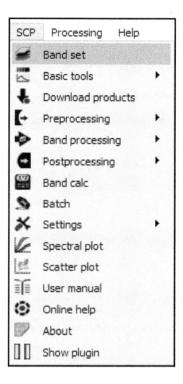

Under **Single band list**, click on the bands that we want to work on (Band 5, 4, and 3) and then click on the + under **Single band list** to create a **Band set**. Now, use the arrow under the **Band set** definition to put **Band 5** first, followed by **Band 4**, and then **Band 3**. Click on **Quick wavelength settings** and then select **Landsat 8 OLI (bands 2, 3, 4, 5, 6, 7)**, tick **Create raster of band set (stack bands)**, and then click on **Run**:

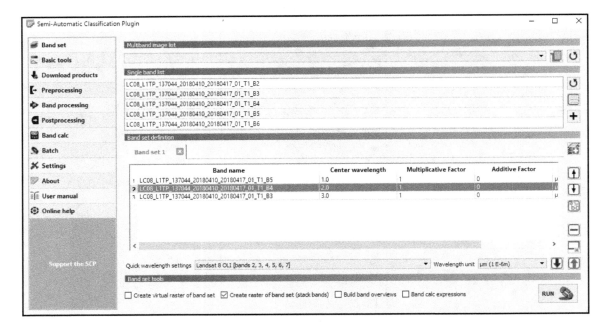

Now, inside **SCP &Dock**, click on **Training input** and then click on **Create a new training input box** (the box with the yellow symbol at the top):

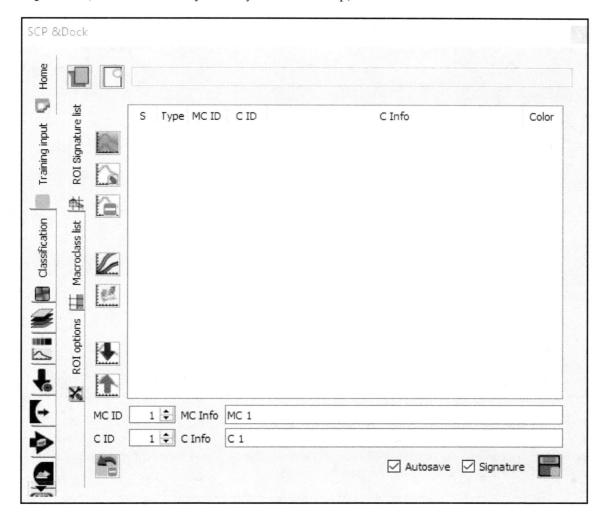

This now creates an SCP file; the training data that we will create now will be populated into this SCP file:

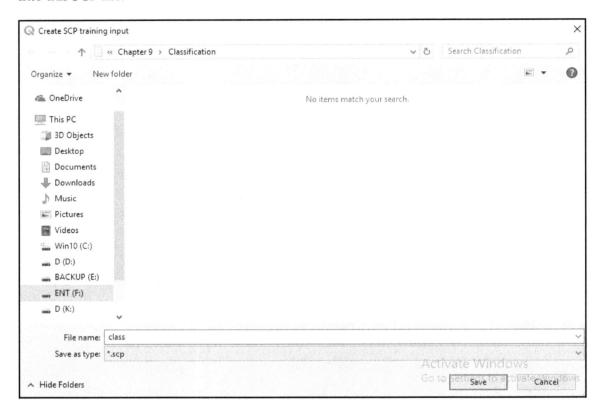

Now, we need to create a training file, and we do so by selecting different areas of the raster file and classifying them according to our judgement or knowledge. Now, to create polygons that will be assigned to different classes, click on the symbol third from the right to start digitizing, and generate the training data:

Start making a polygon by left-clicking, then right-click when the polygon covers the desired area sufficiently. Here we are digitizing to create a signature for the river:

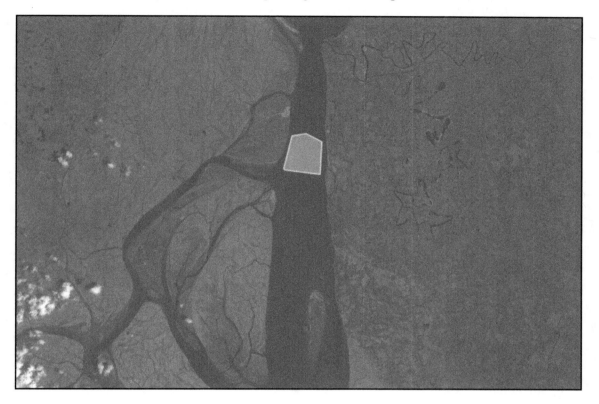

Now, in the **SCP &Dock** window, for **MC Info** (short for **macro class**) and for **C info** (short for **class**) , enter Water in both fields. For **MC ID** and **C ID**, enter 1 in both fields. Now, click on **Signature** to calculate the signature for the Water polygon:

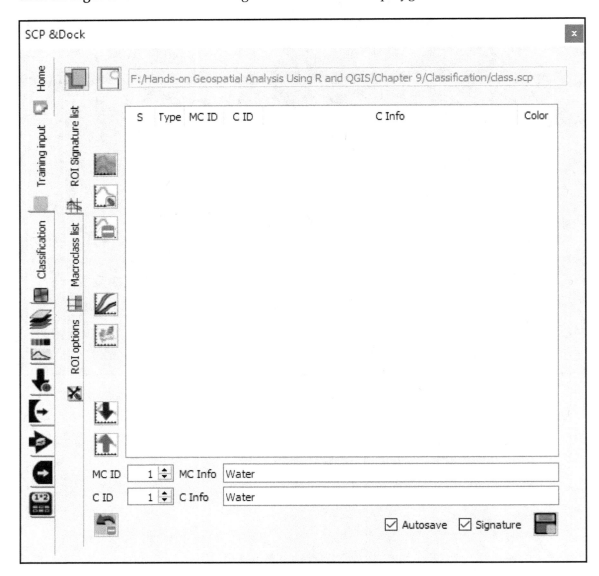

Similarly, do this for other areas of `Water`; that is, make a couple more such polygons of `Water`. Now we will create polygons for `Vegetation`. Select a dark red area and create a polygon as previously outlined:

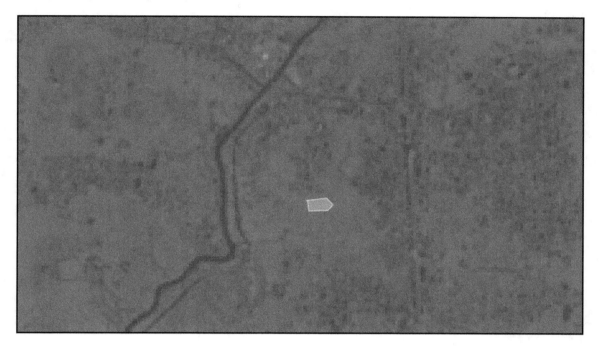

Similar to how we did with `Water`, write `Vegetation` for **MC Info** and **C Info** and write 2 for **MC ID** and **C ID**, and then click on **Signature**:

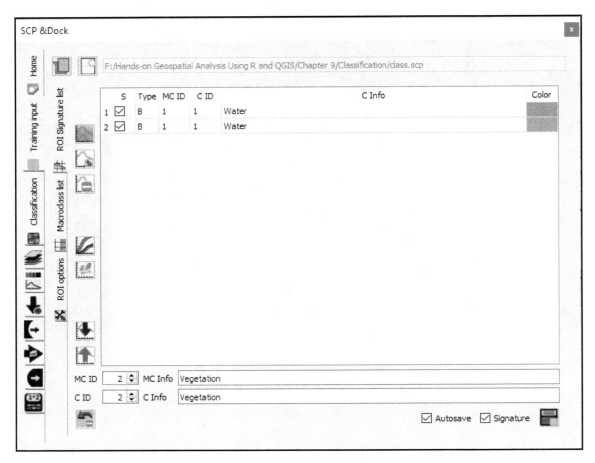

For soil, move to a tan area and make a polygon there:

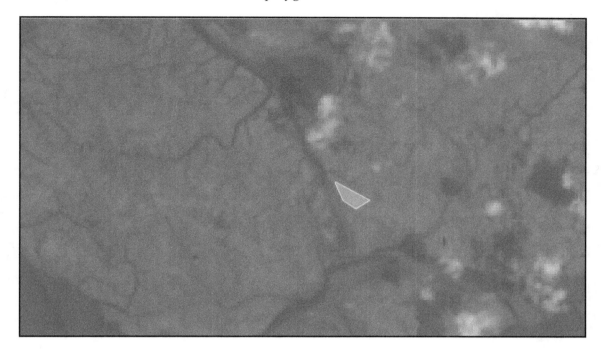

Similar to how we did for **Water** and **Vegetation**, create a signature for **Soil**:

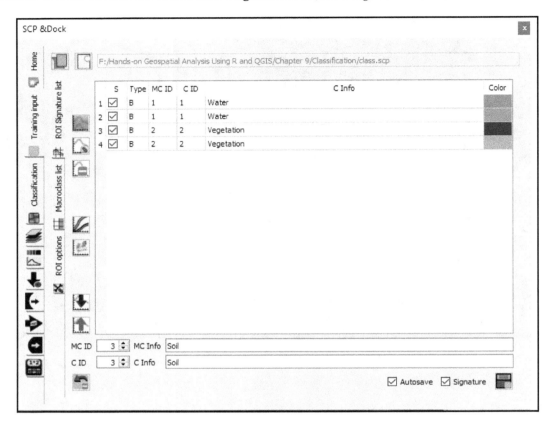

If we find that we have wrongly classified or wrongly generated a signature for a classification, we can remove that item by clicking on the - sign in the bottom-left corner of this screenshot:

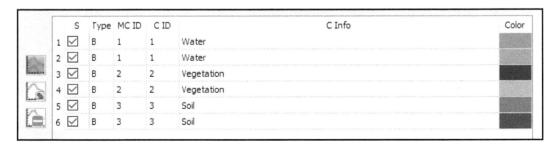

If we want to get the average signature for a class, we can select all relevant signatures (for example, **Water**) by clicking on all signatures of **Water** and then clicking on Merge highlighted spectral signatures obtaining the average signature:

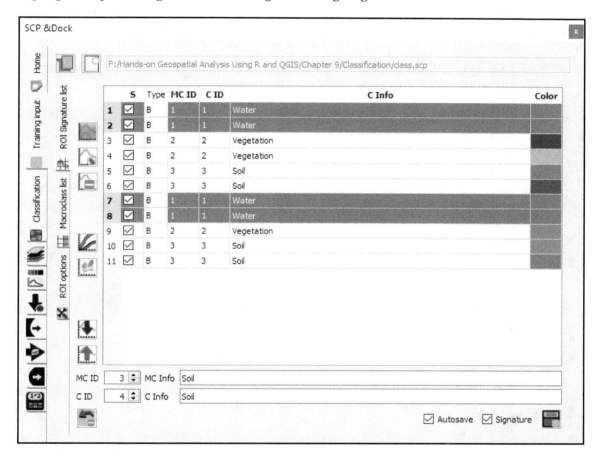

The preceding input results in another signature called **merged_Water**. We can also do this for **Vegetation**, for which we will again get **merged_Vegetation**. After we do the same for **Soil**, also, we get **merged_Soil**:

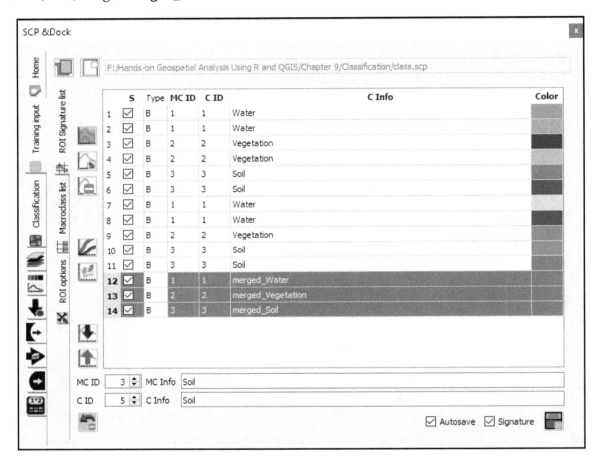

Now, we can click on the plotting symbols to look at how well the spectral separability is. Click on **Classification** and then select **Maximum Likelihood** under **Algorithm**, then tick **Classification report** and click on **OK**:

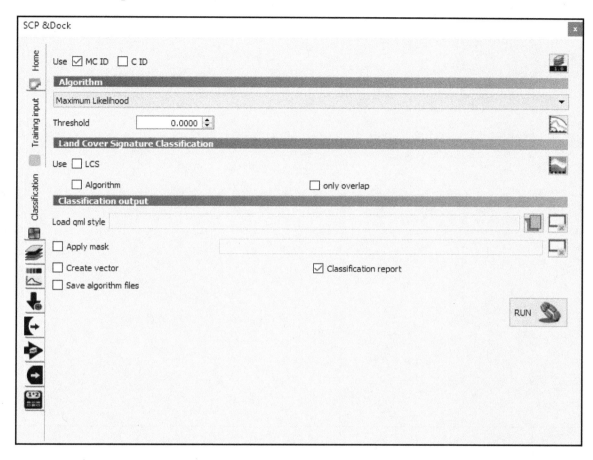

This will now give us an image classified according to the signature that we have generated.

Creating a validation shapefile

We will learn here how to create a validation shapefile. To do so, we need to create a new shapefile and then digitize a raster file to populate the shapefile with attribute information. Now, click on **Layer**, then **Create Layer**, and then on **New Shapefile Layer...**

Now, save it to your desired location and create a new field named **class**, as follows:

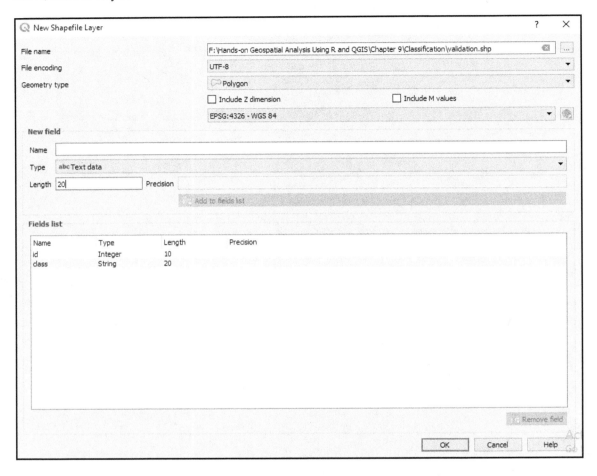

Again, we need to digitize, and so we need to click on **Toggle Editing**:

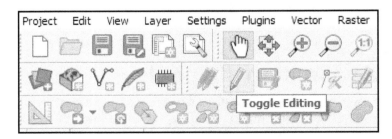

Now, create a polygon by left-clicking and stopping by right-clicking when the desired shape is formed. As soon as we are finished with that, fill in as required:

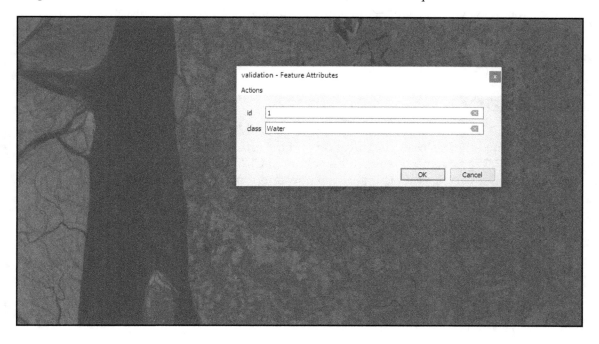

Similarly, do this for **Vegetation** and **Soil** and write their attributes as 2 and 3. Now, if we open the attribute table of this shapefile, we will see something similar to this:

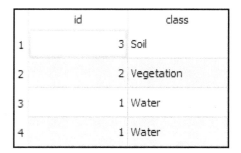

	id	class
1	3	Soil
2	2	Vegetation
3	1	Water
4	1	Water

Unsupervised classification

Unlike supervised classification, in unsupervised classification, the algorithm tries to group or cluster pixels with similar values. In doing so, it tries to group pixels into clusters so that the pixels are homogeneous within clusters and heterogeneous across clusters.

Summary

Here we have learned how to conduct supervised classification with remote-sensing data. We have learned the basics of classification and the implementation details of this in QGIS using the SCP. We have used landsat images and a CIR image to do so. Land cover classification is a very useful technique to know and it must be remembered that we have used only one type of band combination for classification; there are many other band combinations of landsat that can generate useful results for different objectives, such as studying urban areas or vegetation study. In the next chapter, we will use both QGIS and R for landslide susceptibility mapping using the knowledge and skill we have gained so far.

10
Landslide Susceptibility Mapping

In this chapter, we will work with landslide data from Bangladesh and will create a landslide susceptibility map. We will use both R and QGIS to accomplish this. Using a logistic regression model, we will get the prediction probability of a landslide. The basics of model evaluation will also be discussed, and you will be introduced to the machine learning approach to modeling.

The following topics will be covered in this chapter:

- Preprocessing data in QGIS
- Further preprocessing data for model building using R
- Fitting logistic regression in R for predicting landslide and for susceptibility mapping
- Using **Classification and Regression Tree** (**CART**) and random forest for improving model accuracy

Landslides in Bangladesh

Bangladesh is prone to different types of geological, hydrological, and meteorological hazards – landslides are just one of them. Landslides mainly occur in the southeastern part of Bangladesh, which is also called the **Chittagong Hill Tracts** (**CHT**). Landslides mainly occur in monsoon time, and are primarily due to factors such as cutting hills, high slopes, certain degrees of elevation from the sea level, and slope saturation. As people also live in these hilly areas, landslides can be incredibly deadly, as was the case in 2017, when many people died due to multiple landslide events. As such, it is important that we have the ability to model and predict the highly susceptible zones so that lives can be saved and injuries can be prevented.

The **Institute of Remote Sensing, Jahangirnagar University (IRS-JU)**, arranged a field trip to the CHT in November 2017 and took different measures from past landslide locations:

Here we see a raster screenshot of the CHT along with the point locations of the landslide events investigated by IRS-JU. Now, we will use both QGIS and R to make a landslide susceptibility map.

Landslide susceptibility modeling

For landslide susceptibility analysis, we will follow these steps:

1. We will load a **digital elevation model** (**DEM**) from SRTM (provided in the `Data` folder of `Chapter10`).
2. Compute the slope from this DEM.
3. Reproject each file to the same projection system so that we can compare one with another.
4. Extract the elevation and slope values corresponding to the landslide location using the DEM file and the computed slope.
5. Classify areas as safe or unsafe using the range of slopes within which landslides have occurred.
6. Get random points in the safe zone and extract their elevation and slope values.
7. Using the elevation and slope values from both the unsafe and safe locations, fit a logistic regression model.
8. Using the coefficient of the fitted model, make a landslide susceptibility map.

Data preprocessing

Before we start modeling, we need to have all the data regarding the elevation and slope of safe and unsafe locations.

At first, we will load `DEM_PC.tif`, which is the DEM file corresponding to our study area. This appears as follows:

Now, if we right-click on the **DEM_PC** layer and click on **Properties**, we will see that the coordinate reference system is **WGS 84**. We will now convert it to the projected coordinate reference system of **UTM zone 46N**, corresponding to Bangladesh. To reproject, click on **Raster | Projections | Warp (Reproject)...**:

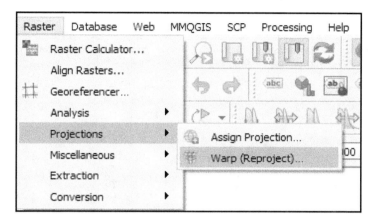

Now, we will get a window for reprojecting. Select **DEM_PC** for **Input layer** and click on the small box to the right of **Target CRS** to reproject:

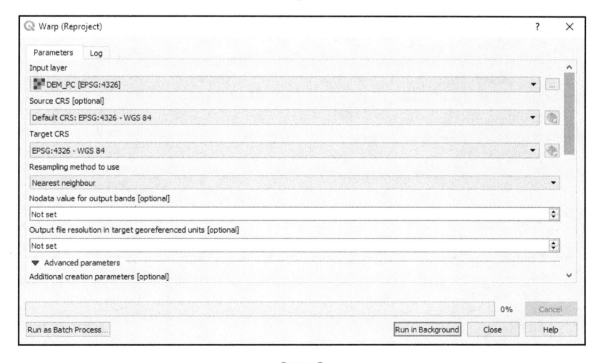

Select **WGS 84 / UTM zone 46N (EPSG:32646)** under **Coordinate reference systems of the world**. Now click **OK**:

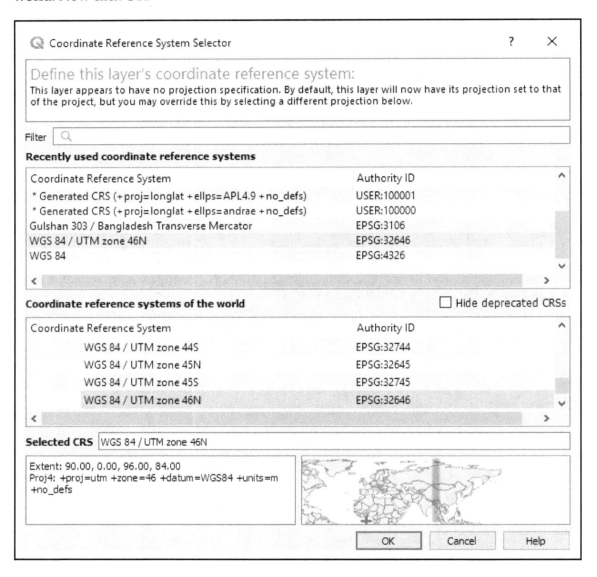

Now, click on **Run in Background**, and after it has run, close it. This will give us the reprojected raster layer of digital elevation. We can save this raster by right-clicking on the layer and then going to **Export** | **Save As...** Under **File name**, browse to the Data folder under Chapter10 and save it as DEM_PC_UTM.tif. Finally, click **OK** to save the raster file:

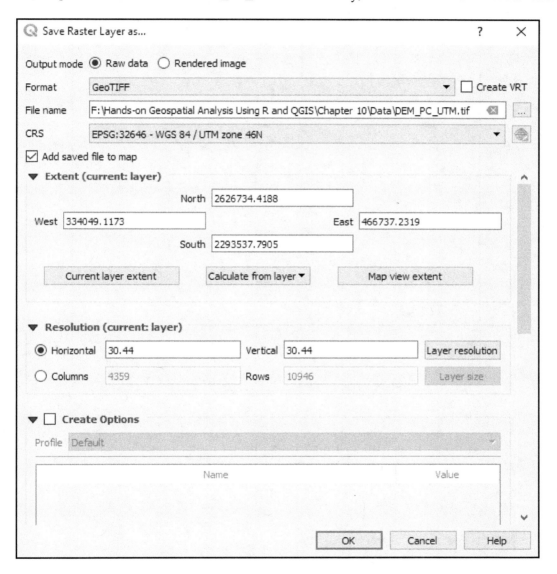

Now, load `DEM_PC_UTM.tif`, which we just created. We need to calculate the slope value from this file for further analysis. We do so by going to **Raster** | **Analysis** | **Slope**. This will look as follows:

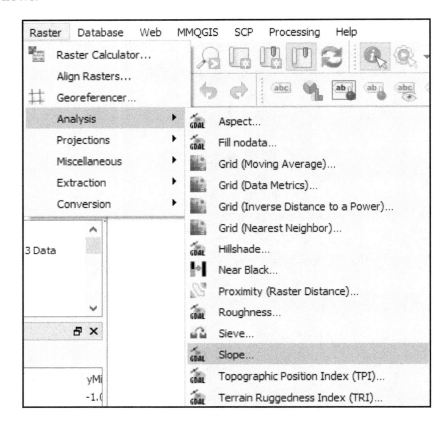

Select **DEM_PC_UTM** for **Input layer** and then click on **Run in Background**:

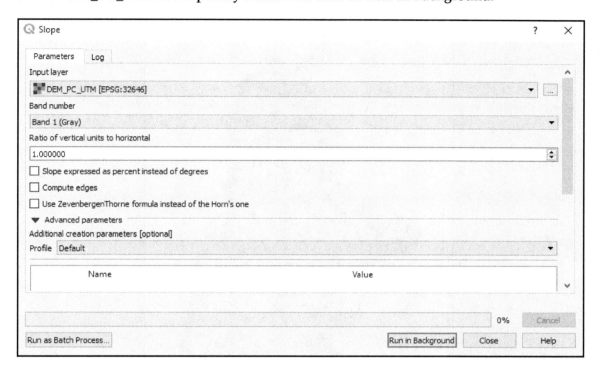

After this command is run, we will get a raster of slope values for the CHT region. Save this as **Slope_PC**:

For a colored version of this or for better visualization, right-click on this raster layer and click on **Properties**. Select **Symbology** in the left panel, then select **Single pseudocolor** under **Render type** and select **Magma** under **Color ramp**, and then click **OK** to visualize the slope in the color magma:

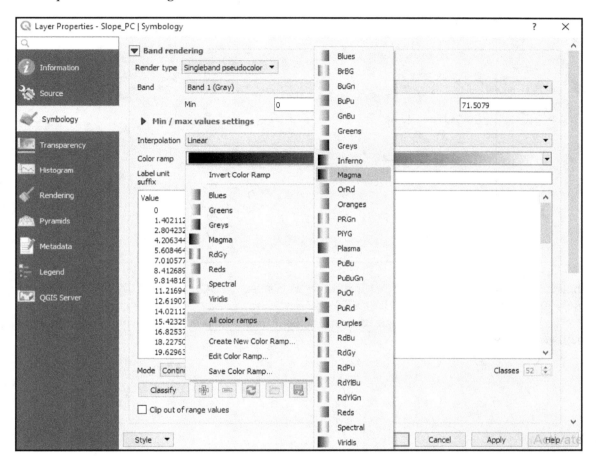

Now, we load the `landslide_locationreal.shp` shapefile from the `Data` folder, and we can check its properties to be sure that its coordinate reference system is **WGS 84**, which we need to convert to **WGS 84 / UTM zone 46N (EPSG:32646)**. Click **Vector | Data Management Tools | Reproject Layer...**:

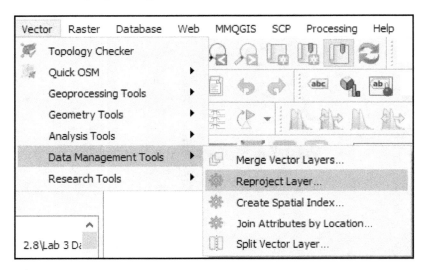

Select **landslide_locationreal** for **Input layer** and select **EPSG:32646 - WGS 84 / UTM zone 46N** for **Target CRS**, then save it as `landslide_location_UTM.shp` by clicking on the box to the right of **Reprojected**. Click **Run in Background**, and after it is finished running, click **Close**:

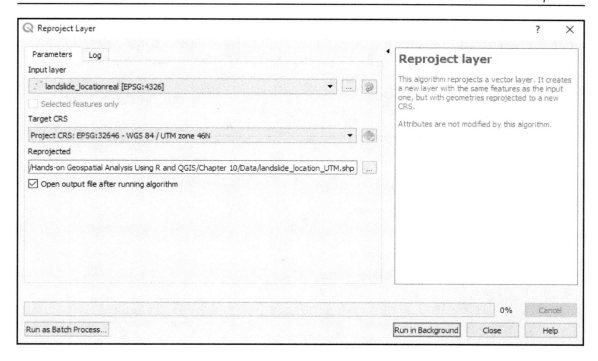

Our multipoint shapefile is now appropriately projected. Now, we need to get the elevation and slope values corresponding to hazard locations. Click on **Plugins** | **Analyses** | **Point sampling tool**. If **Analyses** is not shown under **Plugins**, install it from **Plugins**:

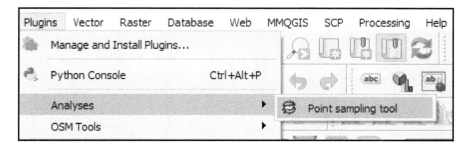

Select **Reprojected** (the reprojected shapefile containing the locations of landslide events) under **Layer containing sampling points**. Highlight **Longitude** and **Latitude** as source points of **Reprojected** and **Slope_PC** as the raster from which the value will be extracted. Under **Output point vector layer:**, browse to the `Data` folder and save it as `slope_hazard.csv`. Click **OK** and it will save the slope values corresponding to the landslide locations in a CSV file named `slope_hazard`:

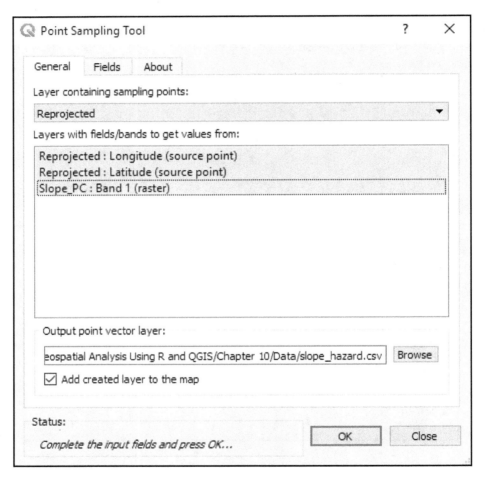

`slope_hazard.csv` will have three columns: `Longitude`, `Latitude`, and `Slope_PC`:

	A	B	C
1	Longitude	Latitude	Slope_PC
2	92.1068	22.5746	10.33463
3	92.2153	22.1879	4.51635
4	91.7879	22.4731	9.0781
5	92.2246	22.1826	3.67018
6	92.2284	22.1986	19.50911

Similarly, we extract values of elevation using `DEM_PC_UTM.tif` and save it as `dem_hazard.csv`.

Now, we will look at the range of the slope values in which landslide occurs. We do so by importing `slope_hazard.csv` in R and by looking at its range:

```
landslide = read.csv("F:/Hands-on Geospatial Analysis Using R and
QGIS/Chapter 10/Data/slope_hazard.csv")
str(landslide)
range(landslide$Slope_PC)
```

We see that the range is between `0.33273` and `23.582223`:

```
[1]  0.33273 23.58223
```

Now, we will classify the area of raster where the value of slope doesn't fall in this range. We will now create a `.txt` file where we will classify values from the minimum to 0.33273 as 0, values from 0.33273 to 23.58223 as `NULL`, and values from 23.58223 to the highest as 0 again. We write the classification values in the `classify.txt` file in the following way:

```
*:0.33273:0
0.33273:23.582223:NULL
23.582223:*:0
```

We will now use **recode** from **GRASS** under **Processing Toolbox** to classify
`slope_PC.tif`. Write `recode` in the search bar under **Processing Toolbox** and double-
click on `r.recode`:

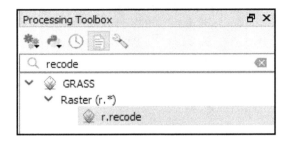

Select **Slope_PC** as **Input layer**, under **File containing recode values**, and browse to
`classify.txt`:

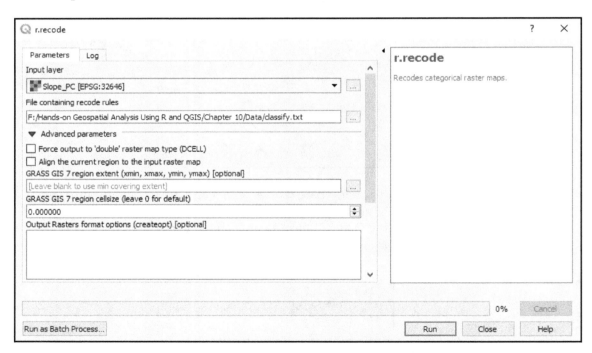

Browse a little further down, click on the box to the right of **Recoded**, and click on **Save to File....** Now click **Run** to get a reclassified raster that has only two values: 0 for safe areas (original slope between 0.33273 and 23.58223) and NULL for other areas:

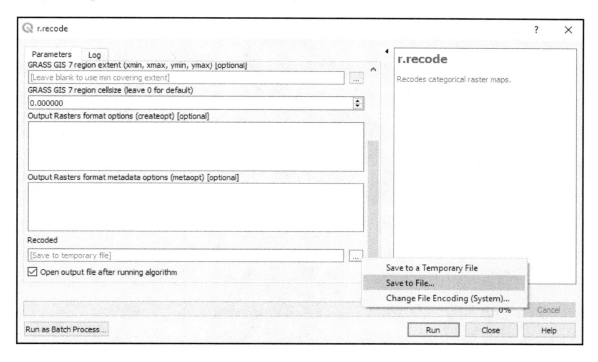

The raster file will now look something like the following screenshot:

Our target is to generate random points within this safe zone. But, before that, we will need to convert the raster to a vector. We do this by clicking **Raster | Conversion | Polygonize (Raster to vector)**:

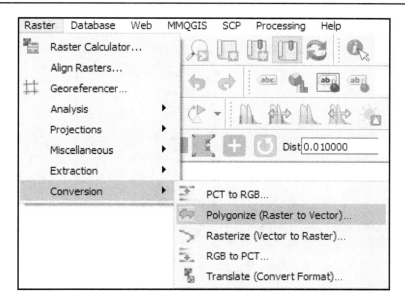

Select **Recoded** as the **Input layer** and then click on **Run in Background** to transform the raster to a vector:

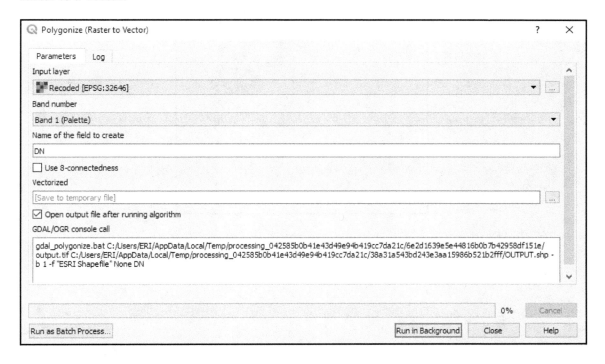

We will now get a polygon of safe zones as follows:

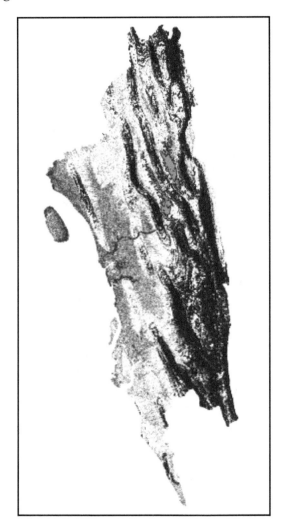

Now save this vector file by right-clicking on the file, clicking on **Export**, and then on **Save Feature As....** Name it `safe.shp` and click on **Save**, which will save `safe.shp`:

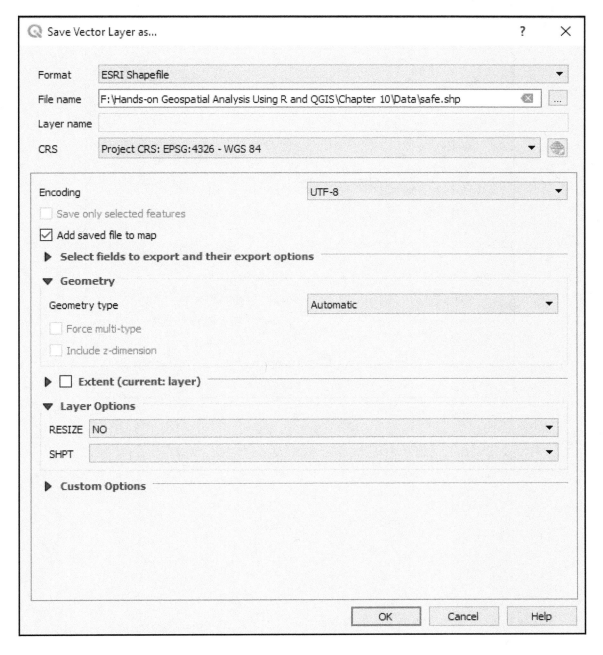

We will create 73 random points (the same as the number of hazard points) in the area covered by `safe.shp`. For those 73 points (in safe zone), we will again record their slope and elevation values. To create random points, click **Vector | Research Tools | Random Points in Layer Bounds...**:

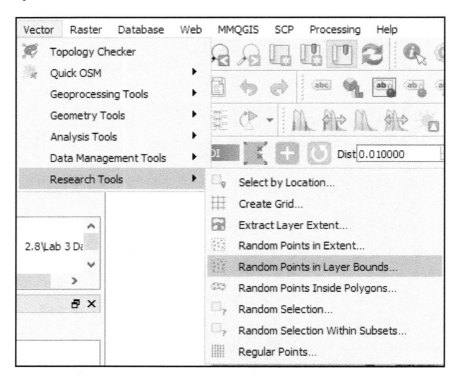

Select **safe** as **Input layer**, write 73 in the value box under **Number of points**, and then click on **Run in Background**:

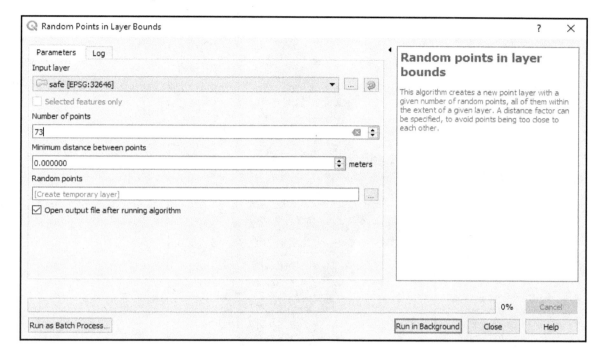

Now we will see that 73 new points are generated inside the **safe** vector. Save it as
`safe_points.shp`:

Now, similar to with hazard location, we first extract the elevation of the points in the safe
zone by clicking **Plugins** | **Analyses** | **Point sampling tool**:

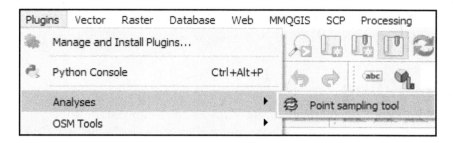

Select **safe_points** under **Layer containing sampling points:** and highlight **safe_points : id (source point)** and **DEM_PC_UTM : Band 1 (raster)**. Save this as dem_safe.csv:

Using the **Point Sampling Tool**, similarly extract the slope values by selecting only **Slope_PC** and **safe_points** and saving it as `slope_safe.csv`.

Model building

Now, we will build our model using all the data we have computed. We want to make a DataFrame of the elevation value, slope value, and an indicator variable indicating whether the location is unsafe or safe. First, load the elevation data corresponding to the safe zone:

```
dem_safe = read.csv("F:/Hands-on Geospatial Analysis Using R and
QGIS/Chapter 10/Data/dem_safe.csv")
str(dem_safe)
# Remove id column
dem_safe$id = NULL
str(dem_safe)
```

By investigating the output, we can be sure that we have, in fact, removed the ID column from `dem_safe`.

Now, load the slope data corresponding to the safe zone and remove the ID column as previously mentioned:

```
slope_safe = read.csv("F:/Hands-on Geospatial Analysis Using R and
QGIS/Chapter 10/Data/slope_safe.csv")
str(slope_safe)
# Remove id column
slope_safe$id = NULL
```

Now, combine a dataset of `dem_safe` and `slope_safe`:

```
safe = cbind(dem_safe, slope_safe)
```

Create a hazard indicator variable, which will have a value of 0 for safe points and a value of 1 for hazardous points:

```
safe$hazard = 0
```

Similar to what we have just done for safe zones, we need to do the same for landslide locations, but now the `hazard` indicator variable will have a value of 1 for all landslide location points:

```
dem_hazard = read.csv("F:/Hands-on Geospatial Analysis Using R and
QGIS/Chapter 10/Data/dem_hazard.csv")
str(dem_hazard)
# Remove Longitude and Latitude column
```

```
dem_hazard$Longitude = NULL
dem_hazard$Latitude = NULL
str(dem_hazard)

slope_hazard = read.csv("F:/Hands-on Geospatial Analysis Using R and
QGIS/Chapter 10/Data/slope_hazard.csv")
str(slope_hazard)
# Remove Longitude and Latitude column
slope_hazard$Longitude = NULL
slope_hazard$Latitude = NULL
str(slope_hazard)

hazard = cbind(dem_hazard, slope_hazard)
# Indicator variable indicating landslide
hazard$hazard = 1
```

Create a composite dataset of `safe` and `hazard` data:

```
landslide = rbind(safe, hazard)
head(landslide)
```

We see that the landslide DataFrame now has all the required values for model building:

	DEM_PC_UTM	Slope_PC	hazard
1	384	25.42593	0
2	110	33.32457	0
3	218	33.63253	0
4	96	28.33403	0
5	115	29.20467	0
6	381	23.96430	0

Now we export this as a `.csv` file for further use:

```
write.csv(landslide, "F:/Hands-on Geospatial Analysis Using R and
QGIS/Chapter 10/Data/model_data.csv")
```

Logistic regression

We will use logistic regression here as it can be used to model dependent variables with values of 1 (for our case, landslide) and 0 (no landslide):

```
logistic_fit = glm(as.factor(hazard) ~ DEM_PC_UTM + Slope_PC,
data=landslide, family=binomial)
summary(logistic_fit)
```

```
Call:
glm(formula = as.factor(hazard) ~ DEM_PC_UTM + Slope_PC, family = binomial,
    data = landslide)

Deviance Residuals:
   Min      1Q   Median      3Q     Max
-2.5078  -0.4583   0.1160   0.5821   1.5567

Coefficients:
             Estimate Std. Error z value Pr(>|z|)
(Intercept)  3.111299   0.489978   6.350 2.15e-10 ***
DEM_PC_UTM  -0.010680   0.005182  -2.061 0.039293 *
Slope_PC    -0.123692   0.033394  -3.704 0.000212 ***
---
Signif. codes:  0 '***' 0.001 '**' 0.01 '*' 0.05 '.' 0.1 ' ' 1

(Dispersion parameter for binomial family taken to be 1)

    Null deviance: 202.40  on 145  degrees of freedom
Residual deviance: 106.62  on 143  degrees of freedom
AIC: 112.62

Number of Fisher Scoring iterations: 6
```

Now, using the estimates of Intercept and coefficients, we calculate the susceptibility. For this case, the value of (Intercept) is 3.111299, the value of DEM_PC_UTM is -0.010680, and the value of Slope_PC is -0.123692. You might find something different for these values.

Now, load the elevation and slope raster:

```
library(raster)
dem = raster("F:/Hands-on Geospatial Analysis Using R and QGIS/Chapter
10/Data/DEM_PC_UTM.tif")
slope = raster("F:/Hands-on Geospatial Analysis Using R and QGIS/Chapter
10/Data/slope_PC.tif")
```

We now calculate the probability or logistic function for all the cells, using the computed coefficients in the following way:

```
val = 3.111299 + dem * (-0.010680) + slope * (-0.123692)
prob = 1/(1+exp(val *(-1)))
```

Here, `rrob` is a raster file that contains the probability of a hazard at each pixel.

Let's get the predicted probability for each observation:

```
logistic_prob = predict(logistic_fit, type="response")
```

Now, create a vector of hazard where we assume that a hazard occurred if the probability is greater than `0.65`:

```
pred_class = rep(0, nrow(landslide))
pred_class[logistic_prob > 0.65] = 1
```

Using a confusion matrix is a way to look at the performance of classification algorithms. It tabulates our classification against reality, the results of which we can use to gain an idea of the usefulness of our model:

		Decision	
		Do not reject H$_o$	**Reject Ho in favor of H$_A$**
Truth	H$_o$ true	True negative	False positive or type 1 error
	H$_A$ true	False negative or type 2 error	True positive

We can create a confusion matrix using `table` in the following way:

```
confusion = table(pred_class, landslide$hazard)
confusion
```

This matrix now looks like this:

```
pred_class  0   1
         0 64   9
         1  9  64
```

Here the accuracy is found by having the total number of correct classifications divided by all classifications. We can compute this from `confusion` in the following way:

```
sum(diag(confusion))/sum(confusion)
```

We should see that the resulting accuracy value is `0.8767`.

Now, let's classify this probability with the following specifications: values between `0` and `0.3` as `0` (say, no risk); values between `0.3` and `0.7` as `1` (say, moderate risk); and values between `0.7` and `1` as `2` (say, high risk). Using `reclassify`, we classify the probabilities into three groups:

```
class = c(0, 0.3, 1, 0.3, 0.7, 2, 0.7, 1, 3)
class_matrix = matrix(class, ncol=3, byrow = TRUE)
risk_class = reclassify(prob, class_matrix)
plot(risk_class)
```

We will now get the following landslide susceptibility map with probability given in three classes:

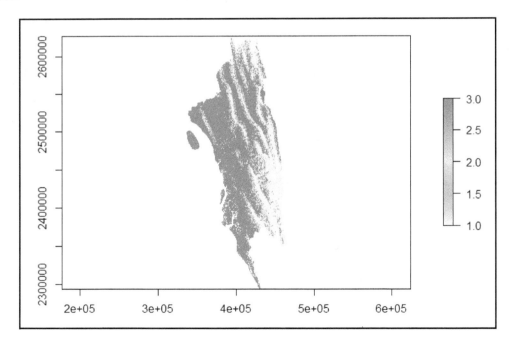

If we want, we can also classify such that we predict a landslide only if the probability is greater than 0.65, otherwise not. We can do so in the following way:

```
prob[prob <= 0.65] = 0
prob[prob > 0.65] = 1
plot(prob)
```

Now, we get the following landslide susceptibility map:

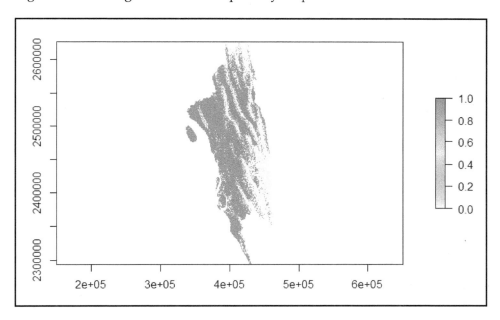

CART

A CART is a classification algorithm that divides the feature space (or independent variable space) into several sections and assigns classes to each subsection. We will not discuss this in detail and will be using R to achieve the classification of areas.

We will now split data into training and testing sets. Training data is normally 70% or 80% of the original dataset that is selected randomly, and this is what's used for fitting the model. After we fit the model, we evaluate our model by fitting it on the remaining 30% or 20% of the data, which we call the test data.

Now, we will import the `landslide` dataset that we exported at the end of our data-preprocessing steps. We will delete the unnecessary `X` column and convert `hazard` into a factor variable for modeling with CART algorithms:

```
landslide_model = read.csv("F:/Hands-on Geospatial Analysis Using R and
QGIS/Chapter 10/Data/model_data.csv")
landslide_model$X = NULL
landslide_model$hazard = as.factor(landslide_model$hazard)
str(landslide_model)
```

Now, install and load all the necessary packages to build a decision tree:

```
install.packages("rpart")
library(rpart)
install.packages("rattle")
library(rattle)
install.packages("rpart.plot")
library(rpart.plot)
```

Now, create training and testing datasets:

```
# Randomly arrange the dataset
set.seed(0)
n = nrow(landslide_model)
random_data = landslide_model[sample(n),]
train = random_data[1:round(0.7 * n),]
test = random_data[(round(0.7 * n) + 1):n,]
```

Now, build a decision tree:

```
tree_train = rpart(hazard ~ ., train, method = "class", control =
rpart.control(cp=0.010))
```

In the preceding code, we use `rpart()` to build a decision tree. In the code, `hazard ~ .`, `train` says that `hazard` depends on all other variables in the training dataset, which are `DEM_PC_UTM` and `slope_UTM`. Then, `method = "class"` says that it is a classification problem and `control = rpart.control(cp = 0.010)` defines the model complexity; the lower the value of `cp`, the more complex or the deeper the tree fits.

We can predict the outcome on the test set in the following way:

```
# Predict the outcome
pred = predict(tree_train, test, type = "class")
```

In the preceding code, we fit the model `tree_train` object on the `test` dataset and we get the predicted class (0 or 1) for each observation in the test dataset.

We can calculate a confusion matrix in the following way:

```
# Calculate the confusion matrix
(confusion = table(test$hazard, pred))
```

By putting the code inside `()`, we execute code and see its content at the same time. The confusion matrix looks as follows:

```
   pred
     0  1
 0  18  0
 1   2 24
```

Here we see that for safe cases (value 0), a safe zone was predicted 18 times and a landslide was never predicted (value 1). For landslide cases, it predicted a landslide 24 times (value 1) and wrongly predicted a safe zone 2 times (value 0).

We can check the accuracy of the model in the following way:

```
sum(diag(confusion))/sum(confusion)
```

We see that the accuracy is higher than that of the logistic regression model now; in fact, it is `0.95`:

```
[1] 0.9545455
```

Random forest

With random forest classification, we build a number of decision trees on training samples (sampling with replacement), and for each split on the tree, a random sample of predictors is chosen. Random forests generally perform very well with classifications. But, for our case, as we have only two predictors, we will not see any improvement. In practical cases, if we took more predictors of landslides, such as the **normalized difference vegetation index (NDVI)**, aspect, and so on, this algorithm would show (for most cases) improved performance.

Similar to what we did with CART, we build the model on the `training` dataset and test its accuracy on the `test` dataset. Now, we build a random forest using the `randomForest()` function of the `randomForest` package:

```
library(randomForest)
rf = randomForest(hazard ~ ., train)
```

`rf` is a random forest object, and we fit `rf` on the `test` data to test its accuracy:

```
pred = predict(rf, test, type = "class")
mean(pred == test$hazard)
```

Using `predict()`, we predict classes on the `test` dataset, and using `mean(pred == test$hazard)`, we get the accuracy of this classifier. As before, the accuracy is still `0.95`.

Summary

In this chapter, we have learned how to model landslide data to predict the likelihood of landslides at different locations and also to produce a landslide susceptibility map. We used QGIS mainly for data preprocessing, and R for modeling. We have used a very simple model and considered only elevation and slope as explanatory variables for landslide. But, as we discussed at the start of this chapter, landslides can depend upon a number of other factors, such as human settlement, vegetation, settlement, tribal agricultural systems (*Jhum*, for Bangladesh), proximity to a drainage system, soil type, and so on. But, using the same technique as the one shown here, you can build a more inclusive model. Furthermore, we have considered only the logistic regression model here; using the same data preprocessing technique, you can now fit more sophisticated models, including neural networks.

Other Books You May Enjoy

If you enjoyed this book, you may be interested in these other books by Packt:

Mastering Geospatial Analysis with Python
Paul Crickard, Silas Toms, Et al

ISBN: 978-1-78829-333-4

- Manage code libraries and abstract geospatial analysis techniques using Python 3.
- Explore popular code libraries that perform specific tasks for geospatial analysis.
- Utilize code libraries for data conversion, data management, web maps, and REST API creation.
- Learn techniques related to processing geospatial data in the cloud.
- Leverage features of Python 3 with geospatial databases such as PostGIS, SQL Server, and Spatialite.

PostGIS Cookbook - Second Edition
Paolo Corti, Stephen Vincent Mather, Et al

ISBN: 978-1-78829-932-9

- Import and export geographic data from the PostGIS database using the available tools
- Structure spatial data using the functionality provided by a combination of PostgreSQL and PostGIS
- Work with a set of PostGIS functions to perform basic and advanced vector analyses
- Connect PostGIS with Python
- Learn to use programming frameworks around PostGIS
- Maintain, optimize, and fine-tune spatial data for long-term viability
- Explore the 3D capabilities of PostGIS, including LiDAR point clouds and point clouds derived from Structure from Motion (SfM) techniques
- Distribute 3D models through the Web using the X3D standard
- Use PostGIS to develop powerful GIS web applications using Open Geospatial Consortium web standards
- Master PostGIS Raster

Leave a review - let other readers know what you think

Please share your thoughts on this book with others by leaving a review on the site that you bought it from. If you purchased the book from Amazon, please leave us an honest review on this book's Amazon page. This is vital so that other potential readers can see and use your unbiased opinion to make purchasing decisions, we can understand what our customers think about our products, and our authors can see your feedback on the title that they have worked with Packt to create. It will only take a few minutes of your time, but is valuable to other potential customers, our authors, and Packt. Thank you!

Index

O

OpenStreetMap (OSM) 45

P

planar point pattern (ppp) 196
point data, geographic information system (GIS)
  adding, on polygon data 54
  importing, from Excel 48
  lines data, plotting in R programming language
    49, 51, 52
  plotting 47, 48
  polygons data, plotting in R programming
    language 49, 51, 52
point pattern analysis, terminology
  events 196
  marks 196
  points 196
  spatial point pattern 196
  spatial point process 196
  window 196
point pattern analysis
  about 196, 204
  G-function 205, 206
  K-function 206, 207
  L-function 208, 209
  marked point patterns 200, 202, 203
  ppp object 197, 199
  ppp object, creating from CSV file 199
  Quadrat test 204, 205
  spatial segregation, for bivariate marked point
    pattern 209, 210
ppp object
  about 197, 199
  creating, from CSV file 199
projected coordination systems 46

Q

QGIS
  environment, obtaining 33, 34, 35, 36, 37, 38,
    39, 40, 41, 42
  installing 32, 33
  URL, for installing 33
  vector data, working 120
Quadrat test 204, 205

R

R programming language
  apply family 20
  area calculation 120
  clipping 116, 117
  data structures 10
  data types 10
  difference 118
  functions 20, 21
  installing 8, 9
  looping 20
  plotting 27, 28, 29, 30, 31, 32
  shapefiles, combining 114
  variable 10
R software
  URL, for downloading 8
range 231
raster data
  aspect 156
  aspect, in QGIS 187
  classification 275
  clip raster, by mask layer 174, 176
  creating, by digitizing 104, 105, 106, 107, 108
  false color composite 155, 163, 166
  hillshade 156
  hillshade, in QGIS 187
  projection system 176
  projection system, changing 178
  projection system, changing of raster file 155
  raster mosaic 169, 172
  rasters, reclassifying 185
  reading 152
  sampling, points used 181
  slope 156
  slope, in QGIS 187
  stacking 153
  working with, in QGIS 163
  working with, in R programming language 151
raster package 8
region 242
remote sensing
  about 149
  atmospheric correction 151
RQGIS package 8

www.ingramcontent.com/pod-product-compliance
Lightning Source LLC
LaVergne TN
LVHW081514050326
832903LV00025B/1491

* 9 7 8 1 7 8 8 9 9 1 6 7 4 *